AMERICAN NATURE GUIDES

MUSHROOMS
AND OTHER FUNGI

AMERICAN NATURE GUIDES

MUSHROOMS
AND
OTHER FUNGI

GEOFFREY KIBBY

SMITHMARK

This edition first published in 1992 by
SMITHMARK Publishers Inc.,
112 Madison Avenue, New York 10016

Published in England by Dragon's World Ltd,
Limpsfield and London

Editor: Diana Steedman
Designer: Mel Raymond
Editorial Director: Pippa Rubinstein
Art Director: Dave Allen

SMITHMARK Books are available for bulk purchase for sales
promotions and premium use.
For details write or telephone the Manager of Special Sales,
SMITHMARK Publishers, Inc.,
112 Madison Avenue,
New York, New York 10016. (212) 532-6600.

ISBN 0 8317 6970 X

Printed in Singapore

Contents

Introduction

North America has one of the richest mushroom floras in the world. Estimates of the number of species that occur are difficult to make since there is still a great deal of work to be done and many species await discovery, but the number is in the thousands, and a pocket guide book cannot hope to show you them all. What are illustrated and described here are over 420 of the most distinctive, important, or otherwise easily identifiable species likely to find, and those mushrooms which are important edibles or are poisonous to some degree.

What is a Mushroom?

Mushrooms are any of the larger fungi which you may find on a walk through the woods, fields, or even in your own backyard. The words mushroom and toadstool, both commonly used, mean exactly the same thing and give no guide as to the edibility or otherwise of the fungus. They have no scientific meaning and are simply names used for several hundred years in many ways by different people in different countries. The larger fungi (anything over about ¼in/0.6cm) are the fruiting bodies or reproductive bodies formed to produce and distribute their spores. The bulk of the fungus consists of almost invisible threads called mycelium, which runs through the soil, wood or other substrate on which it grows and feeds.

The spores are formed in two basic ways. In the majority of the fungi shown in this book – the Basidiomycetes – they are formed on the outside of club-like cells called basidia. In the cup-fungi and related forms – the Ascomycetes – the spores are produced inside long cells called asci and fired out like bullets.

A glance at the colored pictures in this book will show that mushrooms come in a bewildering variety of shapes and colors, and that some method must be adopted in to order to increase your chance of identifying your collections.

How to Use this Book

Firstly you should become familiar with the different parts of the mushrooms you find, and with a very few technical terms used to describe those parts. The pictorial keys show typical mushroom forms and their various parts – in particular the area where the spores are formed, the **hymenium**. This may be on flaps called gills or lamellae, lining the interior of tiny tubes, on spines or teeth, or on an irregular, smooth to wrinkled surface. If your mushroom has gills it is very important to note how those gills join to the stem. They may be completely **free** of the stem; **adnate**, where they join the stem for part or all of their width; **decurrent**, where the gills meet the stem and run down it for some distance; or **sinuate**, where the gill curves up or is notched as it touches the stem.

Make notes on the other features of your mushroom. Is there a ring present on the stem? Does it have a sac or volva at the base? Does the mushroom change color if you bruise it or cut it open. Is there a distinctive odor or taste present? To taste a mushroom(which is very important in some groups) put only the smallest piece on the tip of the tongue and roll it around, never swallow it. If done carefully there is no risk, although it is usual to learn the poisonous species by sight as soon as possible so as to avoid unnecessary tasting of these.

The most important character is the color of the spores in a deposit and to obtain this you must make a spore-print. This is easy to do; simply place a mushroom cap with the gills or tubes or other spore-producing surface down on a sheet of white paper and cover with a cup or other container for a few hours. At the end of this period carefully lift the mushroom cap and underneath you should have a good deposit of spores.

The deposit should be left to dry for a few minutes and then scraped together to observe the color. This will fall into one of four broad categories. The largest is the white to pale-colored group, including pale creams, yellows and ochres to lilac or even greenish. Then there are the brown spores which vary from bright rust-brown through ochre-brown and a dull earthy-brown often called cigar-brown. The third group is the black, blackish-brown, purplish-brown to purple group, and finally there are the pink, or deep salmon-colored spores.

The Descriptions

Once you think you have found an illustration which matches your specimen closely then read the description carefully. All the features should match your specimen – if not then remember there are many more species than can be shown here, and you may have a closely related species. Is your specimen young and in good condition? Old or poorly collected material is difficult to identify.

Microscopic details of the spores are given for each species giving the range of the length and breadth, i.e. spores 7–8 x 4–5μm, which means the spores vary from 7 to 8 micrometers in length by 4 to 5 micrometers in width. A micrometer (often called a micron) is one thousandth of a millimeter.

You do not have to have a microscope to enjoy collecting and identifying mushrooms, but access to one will greatly increase the number of species you can name, and is a fascinating hobby in itself. Some basic chemical reactions are given in some descriptions and if you can obtain some of these chemicals then they can be very helpful. Some are everyday chemicals available in most households. Ammonia for example is present in many household glass cleaners and these are quite adequate for testing. Others are more specialized and their formulae are given in the Appendix.

Collecting

Only collect what you need to study, never over collect and try not to disturb the habitat. The fungus mycelium will remain undisturbed and so picking the mushroom will not endanger the fungus. Always collect all of the mushroom, especially the base of the stem which is very important in some of the poisonous groups. Wrap the specimen in a sheet of waxed paper, rolling it carefully and twisting the ends rather like a candy; this will stop the fungus from being crushed and will retain moisture. Never collect in a plastic bag! Mushrooms soon sweat and become a soggy mess in such bags.

Join your local mushroom group and go on their forays when you can learn from more experienced collectors and experts. There are groups all over the country and the "umbrella" organization, the North American Mycological Association, can put you in touch with one in your area and provide many other valuable services and publications. Their address is given in the Appendix.

Edibility

Because the eating of wild mushrooms is so popular in North America, I have tried to indicate the edibility of each species wherever this is known or has been indicated in other specialist literature. However I must stress that individual reactions to any food can vary, even to well-known edibles, and the author and publishers cannot assume responsibility for the consequences of readers eating wild mushrooms. Do not eat any mushroom without first getting expert opinion on the identification.

Some mushrooms are dangerously poisonous or even deadly, and this has been indicated in the text with a skull and crossbones symbol by the illustration. Many people are poisoned each year, so do not take unnecessary risks. Join your local mushroom society if you wish to learn about eating mushrooms.

Mushroom Names

The majority of mushrooms do not have common names, only those which are well-known edibles or are poisonous, or have some other distinctive feature which caused our forbears to christen them. This often comes as a surprise to people who are used to wild flowers or birds, which nearly all have common names. Fungi, however, have always been rather mysterious or too poorly known to merit such names.

The scientific name of the mushroom consists of two Latinized words followed by the names of the authors who played a part in describing the particular species concerned. For example, if we look at *Tylopilus ballouii* (Peck) Singer we see that the name is Latinized and always in italics. The first part (*Tylopilus*) is the genus or group of mushrooms to which it belongs, and the second (*ballouii*) is the specific epithet, which is always in lower case letters and shows the particular species of the genus we are dealing with. Only one member of the genus *Tylopilus* may bear the specific name *ballouii*.

The author names following the mushroom name show the history of the species: who first described it and whether anyone has ever moved the mushroom to another genus (this happens to many species over the years). In this example the species was first described by Charles Peck (he described it as *Boletus ballouii*), but it was later transferred to the genus *Tylopilus* by Rolf Singer.

Another example is *Russula lutea* (Huds. ex Fr.) S.F. Gray. Notice that the names are usually abbreviated. In this case Fries described the mushroom based on the concept of Hudson, but it was Gray who placed it in the genus *Russula*. Since names do change and vary from guide book to guide book I have given well-known synonyms wherever possible. Remember, the mushrooms do not change, just our ideas about them!

Pictorial Key to Major Groups

Mushrooms with gills:
Both with and without a stem –
see pages 35–134.

Boletes:
Fleshy fungi with spongy tubes on underside of cap –
see pages 16–34.

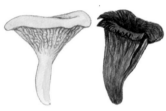

Chanterelles:
Fungi with blunt, irregular wrinkles or veins on underside of cap, or a smooth trumpet –
see pages 150–153.

Puffballs, Earthstars, and Bird's Nests:
Fungi with spores inside a rounded ball, or forming tiny eggs in a "nest" –
see pages 141–149.

Stinkhorns:
Fungi with foul-smelling spores smeared over strange fruit-bodies which hatch out of "eggs" –
see pages 136–140.

Toothed Fungi:
Fungi with downward pointing teeth of spines, with or without a stem – see pages 160–163

Polypores, Bracket Fungi, Crust Fungi:
Hard or tough fungi often with one or more layers of downward pointing tubes – see pages 159, 164–172.

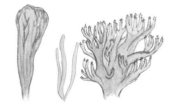

Club and Coral Fungi:
Forming simple clubs or complex coral-like forms – see pages 154–158.
See also 184–185.

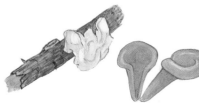

Jelly Fungi:
Very variable in form but all with soft, jelly-like or rubbery texture – see pages 173–174.

Morels, False Morels and Cup Fungi:
From simple cups to sponge- or brain-like structures on a hollow stem. All with spores formed in an ascus – see pages 176–184.

Development of a Mushroom with Volva and Ring

The universal veil (1) enclosing the mushroom ruptures to leave a volva (4) at the base and fragments on the cap. The partial veil (3) covers the gills (2) and then is pulled away to form a ring on the stem.

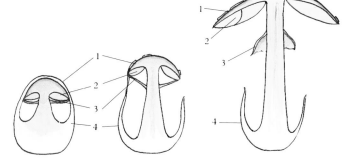

Types of Gill Attachment to the Stem and Cap Shapes

Free gills on a rounded or convex cap

Sinuate or notched gills on an umbonate cap

Adnate gills on a depressed cap

Decurrent gills on a funnel-shaped cap

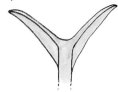

Common Mushroom Structures

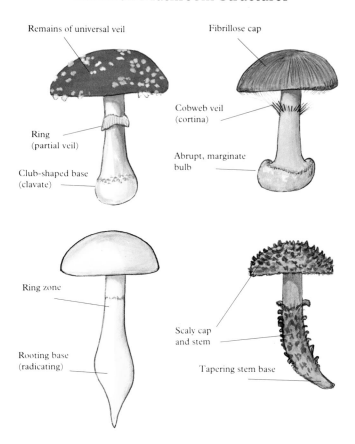

Remains of universal veil

Fibrillose cap

Cobweb veil (cortina)

Ring (partial veil)

Abrupt, marginate bulb

Club-shaped base (clavate)

Ring zone

Scaly cap and stem

Rooting base (radicating)

Tapering stem base

Spore Types

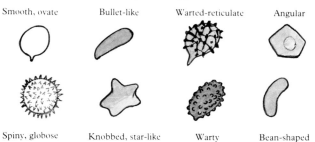

Smooth, ovate

Bullet-like

Warted-reticulate

Angular

Spiny, globose

Knobbed, star-like

Warty

Bean-shaped

Glossary

Adnate Refers to gills which join the stem for most of their width.

Amyloid A blue-black reaction of cells or spores to the iodine solution, Melzer's Solution.

Basidia Club-like cells on which the spores are produced.

Caespitose Growing in clusters or clumps.

Clavate Club-shaped.

Cuticle Outer skin of the cap or stem.

Cystidium Specialized cells on any part of the mushroom, often very distinctive in shape.

Decurrent Refers to gills which run down the stem where they join.

Dextrinoid A reddish-brown reaction in cells or spores to the iodine solution, Melzer's Solution.

Fibrillose With fine fibers.

Fibrous With coarse fibers.

Floccose Wooly or with small tufts of fibers.

Globose Roughly spherical.

Hygrophanous Changing color as it dries out.

Hymenium A layer of cells which produce spores.

Melzer's Solution A specialized iodine solution. See Appendix for formula.

Mycelium The mass of fine threads (hyphae) which form the fungus running through the soil.

Pruinose Frosted or with a "bloom" – like a grape.

Reticulum A fine network.

Sinuate Refers to gills which curve up or are notched where they join the stem.

Striate With fine lines.

Tomentose Finely velvety.

μm, micrometer or micron $1/1000$ of a millimeter

Umbilicate With a central depression or navel in the cap.

Umbo A central hump or nipple in the cap.

Vinaceous Wine-red, purplish-red.

Viscid Sticky, glutinous when wet.

BASIDIOMYCETES

BOLETES

GILLED MUSHROOMS

GASTEROMYCETES

CHANTERELLES

CORAL AND CLUB FUNGI

CRUST AND TOOTHED FUNGI

POLYPORES

JELLY FUNGI

Fungi with spores produced externally on cells
called basidia. These basidia may be spread over
gills, tubes, spines, or other external surface or
contained inside the fruit-body.

BOLETES

These fungi include some of the most sought after
edible species in the world. Mushrooms such as *Boletus
edulis* are eaten all over the world either fresh, dried or in
mushroom soups. They share the common feature of a
soft, fleshy body, the basidia produced in a vertically
arranged layer of minute tubes (the pores) on the
underside of the cap, and most if not all species are
mycorrhizal with certain tree species. Spores vary from
round to more commonly long, elliptic or bullet-like,
smooth to ornamented and range in color from
yellowish to dull olive-brown, pinkish or black.

Other features to note are any color changes, when
the flesh or pores are bruised and occasionally in
chemical tests, or taste and odor. In size they range from
less than 1in to over 2ft (2.5–60cm) across the cap in
some tropical species.

Boletus edulis Cep, or Steinpiltz
Bull. ex Fries
Cap 3–10in/8–25cm. One of the most widely sought after
edible species, it occurs over all northern temperate regions
of the world. It is found in mixed conifer and deciduous
forests. The young pores are white then turn yellow-olive.
The flesh is unchanging when cut. Cap color varies from
yellow-brown to deep reddish-brown. The pale stem is
covered in a fine network of white ridges. **Spores** 13–19 x
4–6.5μm, olive-brown.

Boletus aereus
Bull. ex Fries

Cap 3–8in/8–20cm. Found in
California under tanoak, madrone or
live oak, a rather uncommon
mushroom originally described from
Europe. The deep chocolate-brown
cap, often blotched paler where leaves
have covered it, and the brownish
stem with darker brown reticulation
are distinctive. The young pores are
white then yellowish. The flesh is white.
Spores 12–14 x 4–5μm, olive-brown.
Delicious and much sought after.

Boletus barrowsii
Smith

Cap 3–10in/8–25cm. This large
bolete has an all-white to pale buff
cap and stem with pallid pores. The
stem has a fine reticulum over the
upper half. Cap surface is dry and
slightly velvety. The flesh is white and
unchanging in color when cut.
Spores 13–15 x 4–5μm, olive-brown.
It is often numerous under oaks and pines
in western North America and Mexico. One
of the best edible species rivaling *B. edulis.*

Boletus seperans
Peck

Cap 2–6in/5–15cm. The cap is dull
reddish to liver-brown, paler at the
margin. Pores white then yellowish.
The stem is often a lovely pale wine
color to almost purple, and finely
reticulate. A drop of household
ammonia placed on the stem instantly
turns it sea-green; this separates it from
B. edulis and other similar species. The
flesh is white and unchanging to very pale
blue when cut. **Spores** 11–15 x 3–5μm,
olive-brown. Rather scarce throughout
north-eastern states under mixed
deciduous trees. Edible and good.

Boletus regius
Krombholtz

Cap 3–8in/7–20cm. The beautiful
rose-red to dark-red cap, yellow pores
which bruise blue, and pale yellow stem
flushed pink are easy to recognize. The
stem has a fine network overall. The flesh
stains pale blue. **Spores** 12–16.5 x
3.5–5µm. Frequent in western states and
sometimes reported from the east under
oaks. Edible but often spoilt by insects.
More commonly found in the north-east is *B. pseudopeckii*, at
first rose, but then with duller, browner cap and very fine
network on upper part of stem, under conifers. Edible.

Boletus appendiculatus
Schaeffer

Cap 4–8in/10–20cm. A good edible
species, it has a dry, cinnamon-brown
to yellow-brown cap, bright yellow
pores and stem with a fine yellow
network over the upper half, the base
may be stained reddish-brown. Both
pores and flesh stain blue. **Spores**
12–15 x 3.5–5µm. Found under oaks
in California and some other western
states. Edible.

Boletus calopus
Fries

Cap 2–8in/5–20cm. This is one of a
number of red-stemmed, bitter tasting
species. Its distinguishing features are the
reddish stem with a white network, yellow-
brown to whitish-olive cap and yellow
pores which bruise blue, as does the
flesh. **Spores** 10–14 x 4–6µm. Under
conifers in western North America.
The very bitter taste renders it inedible.
The similar *B. rubripes* has no reticulum
over its red stem, and also occurs in the west. *B. coniferarum*
has a network but lacks red tints on the stem; frequent in the
Pacific Northwest. In eastern states replaced by *B. inedulis*
which has a slender, red stem with a partial to minute network,
similar cap and taste but smaller spores (9–12 x 3–4µm).

Boletus subvelutipes
Peck

Cap 2–5in/5–13cm. Sometimes
mistaken for *B. calopus*,
this differs in its more orange-brown
cap colors as well as paler, orange-red
pores. The stem often has minute
dark red hairs at the extreme base,
you may need to use a hand lens to
see them. All parts bruise blue.
Spores 12–18 x 4–6µm. Rather
common in eastern North America
in mixed woods. Mildly
poisonous.

Boletus subluridellus
Smith & Thiers

Cap 2–4in/5–10cm. The cap
varies from orange-red when
young to deep blood-red when
mature and easily bruises
blackish-blue. The pores are
equally deep red, while the stem
is yellow with fine red dots, the
base has a pale yellow woolly
coating. The flesh turns deep
blue when cut. **Spores** 11–15 x
4–5.5µm. This beautiful species is
not uncommon in warm, wet
years in late summer under mixed
deciduous trees in eastern North
America. Edibility unknown.

Boletus frostii
Russell

Cap 2–6in/5–15cm. Unmistakable
with its deep apple-red to blood-red
coloration over the entire mushroom,
with a strongly raised network over
the stem, often paler in color. All
parts bruise blue. **Spores** 11–15 x
4–5µm. Occasional to frequent under
oaks in eastern and southern states.
Edible but not recommended. One of
the world's most spectacular fungi.

Boletus satanus
Lenz

Cap 3–12in/7–30cm. This often huge, bulky bolete is easily recognized by the swollen, obese stem with fine red network and the pallid to grayish-olive cap which soon turns pink with age. The blood-red to pinkish pores bruise blue as does the pale yellow flesh. Causes nausea, vomiting and cramps and is, like all boletes with red pores, best avoided. Frequent some years under oaks in California. **Spores** 11–15 x 4–6μm. The American fungus differs markedly from European *B. satanus* which has no pink tones in the cap and a repulsive rotten-garlic odor with age.

Boletus pulcherrimus
Thiers & Halling (=*B. eastwoodiae*)

Cap 3–8in/7–20cm. A striking species with its blood-red pores and red-netted stem, it is not as bulky or fat-stemmed as *B. satanus* and the cap is darker, olive to reddish-brown. Flesh and pores bruise blue. **Spores** 11–16 x 5–6.5μm. Found in mixed woods along the Pacific Coast. Mildly poisonous and should not be eaten.

Boletus pseudosensibilis
Smith & Thiers

Cap 3–6in/8–15cm. Very common in eastern states, it has a brick-red or russet color cap, yellow pores which can be slightly decurrent and bruise deep blue, and a stem flushed bright red. The flesh, when cut, instantly turns deep cobalt-blue; compare this with *B. bicolor* with which it is often confused. Dilute ammonia dropped on the cap turns deep sea-green, distinguishing it from other similar species. **Spores** 9–12 x 3–4μm. Usually under oaks and beech. Edible but best avoided.

Boletus bicolor
Peck

Cap 2–6in/5–15cm. Starting a beautiful rose-red it soon fades to dull tan and may often crack in the sun. The pores bruise deep blue but the flesh hardly turns blue at all. The often pointed stem shades to deep purple-red at the base.
Spores 8–12 x 3.5–5µm. Often abundant in eastern North America usually under oaks. Edible.

Boletus longicurvipes
Snell & Smith

Cap 1–2in/2–5cm. This rather small, slender bolete has a tacky, viscid cap varying from reddish-orange to tawny-brown, and a whitish stem flushed pinkish-brown to dull reddish below. It is ornamented with fine reddish floccules. The flesh is white and unchanging when cut. **Spores** 13–17 x 4–5µm. Often a locally common species in mixed conifer woods in the north-east southward to North Carolina. Edible but not worth eating.

Boletus subglabripes
Peck

Cap 2–4in/5–10cm. The often pitted, lumpy or irregular surface of the cap is dry to slightly tacky and a rich tawny-brown to cinnamon color. The pores are yellow while the yellow stem flushes reddish over the lower half. The surface of the stem is minutely scabrous-floccose. The flesh is pale yellow. **Spores** 11–14 x 3–5µm. A common species throughout the eastern states it grows in mixed woodlands. Edible.

Boletus ornatipes
Peck

Cap 2–6in/5–15cm. Cap starts dark
olive-gray but soon becomes flushed
with yellow, while the pores, tubes,
stem and flesh are all yellow, bruising
orange-brown. The stem is strongly
reticulate. Similar in some respects to
B. griseus also shown here, it differs in its
much stronger yellow tones, especially in
the pores and flesh. **Spores** 9–13 x
3–4μm. A common species along
roadsides, banks and woodland edges
throughout eastern states. Edible.

Boletus griseus
Frost in Peck

Cap 2–6in/5–15cm. A dull, grayish-
brown and slightly felty-fibrillose, the
cap may develop yellowish tones with
age. The pores and tubes are pallid to
grayish. The stem is whitish with a greenish-
yellow flush at the base, eventually it may be
yellowish overall. The surface has a fine
network overall. The flesh is pallid then dark
brownish where bruised. **Spores** 9–12 x
3.5–4μm. A rather common species in open
oak woods throughout eastern states. Edible.

Boletus parasiticus
Fries

Cap 1–3in/2–8cm. This unique
species is to be found attached
to the common Earthball,
Scleroderma citrinum, whose
tissues are invaded by the
bolete. Often 3 or more boletes
can be seen on one Earthball which
does not seem to be affected by the
invading bolete and continues to
produce its own spores! The pores
of the bolete are often stained reddish.
Spores 12–18 x 3.5–5μm. Frequent in
wet seasons in north-east and south.
Edible but not worthwhile.

Boletus pulverulentus
Opatowski

Cap 2–4in/5–10cm. Dull yellow-brown to reddish-brown. Pores are yellow and the yellow stem is flushed with reddish-orange below. The remarkable part of this fungus is the instantaneous change of all parts of the fruit-body to the deepest blue when bruised in any way. **Spores** 11–14 x 4.5–6µm. Quite common in lawns and woodlands, often on banks and hillsides, especially under oaks, throughout north-eastern states. Edible but not recommended.

Boletus badius
Fries

Cap 2–4in/5–10cm. The slightly velvety to smooth cap is a rich bay brown to yellow-brown. The yellowish pores become slowly greenish with age and bruise blue. The stem is colored like the cap with pinkish-brown tones, often with a whitish bloom. The whitish-yellow flesh bruises pale blue. **Spores** 10–14 x 4–5µm. A very common species in both conifer or deciduous woods in the eastern states. Edible and good.

Boletus rubellus
Krombholtz

Cap ½–2in/2–5cm. A small bolete, bright rose-red to scarlet when young with a velvety surface, becoming paler and often cracking with age. The yellow pores and reddish stem both bruise blue when handled. The base of the stem has a distinct yellow coating. **Spores** brown, 7–17 x 4–7µm. Found in grass under oaks, quite common. A number of small red species are distinguished microscopically, such as *B. campestris* and *B. fraternus*, but they are doubtfully distinct.

Boletus mirabilis
Murrill

Cap 2–6in/5–15cm. A deep liver-brown to chocolate bolete with a roughened, almost reticulate surface to the stem, and frequent occurrence on rotted conifer logs makes this western species a very striking fungus. The flesh is pallid and unchanging when cut while the pores are pale greenish-yellow. **Spores** very large, 19–24 x 7–9µm. It is considered an excellent edible species.

Boletus affinis
Peck

Cap 2–4in/5–10cm. Surface dry and slightly roughened, usually a dull, warm brown and often spotted with small decolored areas. The pores are white then pale buff, while the stem is whitish above and reddish-brown below, and is quite smooth. The white flesh is unchanging. **Spores** 12–16 x 3–3.5µm, yellow-brown. A common species throughout eastern states in mixed woodlands, it is regarded as edible.

Boletus piperatus
Fries

Cap 1–3in/3–8cm. A small but attractive species with a dull, orange-brown cap and intense cinnabar-red to cinnamon-colored pores. The stem is pale brown with the base always a bright chrome-yellow, as is the flesh when the stem is split open. **Spores** 9–12 x 4–5µm. As the name suggests the taste is very peppery. Frequent but rarely common, throughout the north-east under pines and birch.

Boletus caespitosus
Peck

Cap 2–3in/5–8cm. The often clustered fruitings, plus the vivid yellow, unchanging pores are distinctive as is the slightly sticky cap when wet. The odor of the flesh when cut is also very strong, exactly like that of the common Earthball (*Scleroderma citrinum*) and is unique in North American boletes. **Spores** 9–10.5 x 4–4.5µm. Frequent in hot, wet summers in eastern states under oak and beech. Edibility is unknown.

Boletus spadiceus
(Fr.) Quel.

Cap 1–4in/3–10cm. The cinnamon to yellowish-tan cap is usually rather velvety and a drop of ammonia produces an instant bright green reaction. The pores are deep yellow and do not turn greenish when bruised. **Spores** 9–14 x 4–5µm. The stem is distinctive with often prominent ridges and lines at the apex. Commonly found under oaks throughout eastern states. The very similar *B. illudens* also turns green with ammonia, but has pores which do not bruise greenish, and a more prominent network on the stem.

Boletus chrysenteron
Fries

Cap 2–3in/5–8cm. One of the most well-known species in both America and Europe, it often has a cracked cap surface with reddish flesh showing through the cracks. Cap is shades of olive-brown to reddish-brown and the pores are yellow, bruising blue. The stem is yellowish-white above shading to purplish-red below. The flesh bruises blue. **Spores** 9–13 x 3.5–5µm. Common in mixed woodlands throughout America. Edible but poor quality.

Boletus ravenelii
(Berk. & Curt.) Murrill
Cap 2–4in/5–10cm. Bright yellow
with a dry, powdery cap surface.
Pores and stem are also yellow and
the stem is joined to the cap margin
by a distinct veil of tissue. The pores
bruise bluish. **Spores** 8–10 x 4–5µm.
This remarkable species is rather
uncommon and grows in pine woods
from the south-east to northern states.
Edible.

Tylopilus alboater
(Schwein.) Murrill
Cap 3–6in/7–15cm. The genus
Tylopilus differs from *Boletus* in its
pinkish-brown to purplish spores and
pink tubes and pores when mature.
This striking species has a velvety,
almost black cap and stem when
young. As it ages the cap becomes
paler, more grayish-black and often
cracks all over. The flesh when cut
turns first reddish-gray then black.
Spores 7–11 x 3.5–5µm. In some
seasons it can be quite common under
oaks in eastern states. Edible, with a
mild taste.

Tylopilus felleus
(Fries) Karsten
Cap 2–8in/5–20cm. Often very large
in size this appears to be a tempting
edible, and is confused with the Cep
bolete, *Boletus edulis*, but one taste
shows the difference, it is terribly
bitter! The cap and stem are
yellowish-brown to tan while the
pores mature to deep pink. The stem
has a raised network over the whole
surface. **Spores** 11–15 x 3–5µm.
Common under mixed pines and
deciduous trees throughout east from
Florida to Maine.

Tylopilus eximius
(Peck) Singer

Cap 2–6in/5–15cm. An unusually colored bolete with chocolate to purplish brown cap, even darker purple-brown pores and pale lavender-purple stem with darker, woolly squamules. The taste is mild, unlike most *Tylopilus* species, and it is usually considered a good edible. **Spores** 11–15 x 3–5µm. Widely distributed from Maine southward to Florida in mixed woodlands but never common.

Tylopilus plumbeoviolaceus
(Snell & Dick) Singer

Cap 2–6in/5–15cm. This can be one of the most beautifully colored boletes when young with a deep violet cap and stem contrasting with white pores. As it ages the mushroom turns paler lavender-violet to gray-violet and the pores mature pink. **Spores** 10–13 x 3–4µm. This is one of the most common boletes in late summer to early fall in deciduous woods throughout eastern states. It is bitter to taste and inedible.

Tylopilus chromapes
(Frost) Smith & Thiers

Cap 2–6in/5–15cm. When fresh this bolete presents one of the most remarkable colorations in the fungus world with its pale strawberry-pink to reddish cap, pink stem with darker red woolly scales and a brilliant yellow base, and white pores turning pale pink. **Spores** 11–17 x 4–5.5µm. It is common in deciduous woods throughout eastern states. The taste is mild and it is a good edible species.

Tylopilus ballouii
(Peck) Singer

Cap 2–4in/5–10cm. The bright, burnt-orange cap is very unusual among boletes. Combined with the pale cream stem which is smooth and ages or bruises browner, and almost white pores which bruise brown, it is usually easy to identify. When old the cap can, however, fade to a pale tan. The taste is slightly bitter. **Spores** 5–11 x 3–5μm. Frequent in warmer eastern states such as southern New Jersey south through Florida, under oaks and beeches.

Boletellus russellii
(Frost) Gil.

Cap 2–6in/5–15cm. The genus *Boletellus* is separated from *Boletus* because the spores of all the species are ornamented with longitudinal ridges, wings or grooves. Many – as in this case – have remarkably shaggy or coarsely reticulated stems, and often shaggy caps as well. The stem is usually very tall in relation to the cap diameter. **Spores** 15–20 x 7–11μm, with longitudinal grooves. A rare find but a spectacular one. It occurs under oaks and pines throughout eastern states. Edible.

Leccinum atrostipitatum
Smith, Thiers & Watling

Cap 3–8in/8–20cm. The genus *Leccinum* is distinguished by the wooly squamulose stem with the squamules often starting paler and darkening with age. With its pale, dull apricot-buff to pinkish cap and large, black-speckled stem, this is a very striking species. The squamules are black even when very young. The flesh when cut slowly turns dull wine-red to violaceous-brown. **Spores** 13–17 x 4–5μm. This species is

frequent under birch throughout eastern and northern states. Reported to cause stomach upsets when eaten.

Leccinum insigne

Smith, Thiers & Watling

Cap 2–6in/5–15cm. The cap is rich orange to brick-red while the woolly squamules on the stem start white but soon age reddish-brown then blackish. The tubes and pores are pale cream bruising brown. The flesh when cut turns grayish-purple and finally purplish-black. **Spores** in both are 13–16 x 4–5μm. Under aspen over much of America. The very similar *L. aurantiacum*, also under aspen, has flesh which turns distinctly reddish before becoming gray to black. Both species are good edibles.

Leccinum rugosiceps

(Peck) Singer

Cap 2–6in/5–15cm. The yellow-orange to tawny cap, which is usually pitted, uneven or wrinkled, and often cracks with age is the key character to look for along with the yellow pores, stem, and slightly darker stem squamules. The flesh turns dull reddish when cut. **Spores** 16–21 x 5–5.5μm. This species is not uncommon in mixed woods especially under oaks throughout eastern states. Considered edible.

Leccinum albellum

(Peck) Singer

Cap 2–3in/5–8cm. This is a very pale, delicate, thin-stemmed species whose entire fruit-body is pallid white to grayish-white. The cap usually cracks into a fine mosaic. The flesh when cut is unchanging. **Spores** 15–20 x 4–6μm. Rather common, especially under oaks, throughout eastern states. The similar *L. holopus* rarely cracks and has flesh turning slightly pink. The white body is often flushed with pale green.

Leccinum scabrum
(Fries) S.F. Gray

Cap 2–6in/5–15cm. Although commonly reported, this is probably the most misidentified species of *Leccinum*. The true *L. scabrum* has a soft, dull brown to buff cap, pale whitish-buff pores that hardly stain when bruised, and stem squamules starting white then soon aging brown to black. The flesh when cut is unchanging to only very slightly pinkish-buff in the stem apex. There may be blue-green stains in the stem base. Other similar species stain a much brighter pinkish-red in the stem apex. **Spores** 15–19 x 5–7μm. A rather uncommon species under birch in the north-east, it is edible but poor.

Gyroporus purpurinus
(Snell) Singer

Cap ½–2in/1–5cm. A small but very attractive species with unusual colors for a bolete, a bright reddish-purple, wine color in both cap and stem. The pores are white to cream. Like all species in the genus *Gyroporus* the flesh is rather hard and brittle, and the stem is usually hollowed out. **Spores** in this genus are shades of yellow and rather short; here they are 8–11 x 5–6.5μm. A rather rare species, it occurs under oaks and other deciduous trees in southern and eastern states. It is edible but hardly worth picking for food.

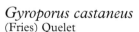

Gyroporus castaneus
(Fries) Quelet

Cap 1–4in/3–10cm. A common species varying in color from light, yellowish-brown to rich cinnamon or brick-red, sometimes with discolored areas. The stem is colored like the cap and is usually hollow. The pores are white to pale yellow. **Spores** 8–12.5 x 5–6μm. A good edible species with a nutty flavor and often abundant under oaks in late summer and fall.

Gyroporus cyanescens
(Fries) Quelet

Cap 2–5in/5–12cm. At first this
species is entirely white but soon ages
a dull yellowish-tan to straw color.
The surface of the cap is roughened
and fibrillose as is the stem which is
usually quite swollen. The most
remarkable feature of this bolete
however is the instant color change to
the most intense deep blue if any part
is damaged. **Spores** 8–10 x 5–6μm. It is infrequently found in
sandy soils under mixed conifer/deciduous trees from Florida
to Maine. Despite its lurid color, it is edible.

Suillus granulatus
(Fries) Kuntze

Cap 2–4in/5–10cm. This has many of
the features considered typical of the
genus *Suillus*: viscid, slimy cap when
wet, stem with small glandular dots
and a strict relationship with pines.
The cap color varies from pale
cinnamon-brown, as shown, to quite
bright orange-brown. The pores are
pale yellow, often weeping milky
droplets when young. **Spores** 7–10 x
2.5–3.5μm, dull brown. It is edible
and widely eaten but as with most *Suillus*
can sometimes cause stomach upsets.

Suillus luteus Slippery Jack
(Fries) S.F. Gray

Cap 2–6in/5–12cm. A common and
attractive bolete with a deep tawny-
brown to reddish-brown, slightly
streaky cap with a glutinous, viscid
coating. The stem is noticeable for its
thick, membranous veil forming a ring
at the top, white and often flushed
with violet on the underside.
Spores 7–9 x 2.5–3μm. A widely
sought after edible species it is
common under pines in eastern
North America.

Suillus grevillei

(Klotzch) Singer [= *S. elegans*]
Cap 2–6in/5–15cm. An often
abundant species, its cap
varies from bright yellow to
deep reddish-chestnut while
the yellowish stem has a
white, cottony ring at the
top. The pores are yellow,
bruising reddish-brown.
Spores 8–10 x 3–4µm.
Edible, but as with other
species it is best to remove
the slime which seems to
cause some people stomach
upsets. Found throughout
most of America, only
under larch.

Suillus tomentosus

(Kauffman) Singer, Snell & Dick
Cap 2–6in/5–15cm. The
specific name refers to the
cap surface which is finely
hairy-velvety to squamulose
at first then becoming
smooth, dry but sticky below
the tomentose layer. The
color varies from pale yellow
to orange-yellow as does the
stem. The pores are dark
brownish-cinnamon
becoming dull yellow with
age. The pores, and the flesh
when cut, turn blue. **Spores**
7–10 x 3–5µm. Under
two-needle pines it is quite
common in western states
and considered edible.

Suillus americanus
(Peck) Snell ex Slipp & Snell
Cap 1–4in/3–10cm. The cap is slightly conical when young, bright yellow frequently blotched with bright red streaks and spots. The cap margin usually has a soft, cottony veil hanging down. The stem has darker brown glandular spots while the pores are rather large, and honeycomb-like. **Spores** 8–11 x 3–4µm. Edible but rather poor. It can be very common, in large circles around white pine throughout eastern states.

Suillus pictus
(Peck) Smith & Thiers
Cap 2–6in/5–15cm. A beautiful species especially when young, it has a bright red, fibrous-scaly cap, quite dry, which fades to reddish-ochre with age. The stem is also red and scaly with a ring-like zone at the apex. The pores are usually decurrent , bright yellow and large. **Spores** 8–11 x 3.5–5µm. Often very common under white pine it is a good edible.

Suillus cavipes
(Opat.) Smith & Thiers
Cap 2–4in/5–10cm. The cap is a dull yellowish-red to reddish-brown, dry, scaly-fibrous. The stem is reddish-brown and usually hollow with a white, ring-like zone at the apex which soon vanishes. The pores are decurrent and dull yellow to greenish-yellow and large, angular. **Spores** 7–10 x 3.5–4µm. Frequent under larch, it is found from Nova Scotia to Washington. Considered a good edible.

Boletinellus meruloides
(Schweinitz) Murrill

Cap 2–4in/5–10cm. Common
wherever ash trees are found,
this is a rather unattractive
bolete with a dull yellow-
brown to red-brown cap with
decurrent pores that are often
rather gill-like and not easily
separated from the cap flesh,
unlike other boletes. All parts
bruise a darker, reddish-
brown. **Spores** 7–10 x
6–7.5μm, olive-brown.
Edible but very poor flavor.

Strobilomyces confusus
Singer

Cap 2–6in/5–15cm. This genus is
particularly common in the tropics
and only 3 or 4 species occur in
North America. All members are
distinguished by their almost black
spores which are often globose
and strongly ornamented. In
eastern America 2 species are
common, *S. confusus* and *S.
floccopus*. The latter is more
usually included in guidebooks
but in fact *S. confusus* is frequently
the more common of the two.
Both share almost identical black,
shaggy caps and stems, white to
gray pores and flesh that turns
orange-red then black. *S. confusus*
however can usually be
distinguished by its narrower,
sharper and much firmer cap
scales. It has spores with only a
partial reticulum compared with
the full network found on *S.
floccopus*. **Spores** 10.5–12.5 x
9.5–10μm. Both are found under
mixed woods from late summer to
fall and both are edible but of
poor quality.

GILLED MUSHROOMS
WHITE TO PALE OCHRE SPORES

RUSSULA FAMILY

These two genera are often placed in their own order –
the Russulales – because they differ so markedly from
other gilled mushrooms. They have a characteristic
brittle, crumbly texture in the hand, and their spores are
ornamented with a variety of warts and ridges that stain
blue-black when treated with a special iodine solution
(Melzer's Solution). Other important features are the
taste of the cap flesh and gills, the odor, and the color of
the spore deposit. The latter is placed on a scale of pure
white to ochre, designated by 8 letters (A–H) shown
below and is vital to accurate identification, especially in
Russula.

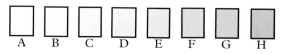

A   B   C   D   E   F   G   H

Lactarius differs from *Russula* in its usually duller
colors and in its flesh which oozes a sticky latex when
cut. This latex may be clear, white or colored. The spore
color is less variable in *Lactarius*. Both groups are very
large with several hundred members and are very
difficult to distinguish. Only the most striking and easily
identified species are shown here.

Russula dissimulans
Shaffer

Cap 2–6in/5–15cm. Soon
somewhat funnel-shaped
the cap starts white but ages
blackish-brown. The cap cuticle
(skin) does not peel off easily. The
flesh stains strongly reddish then
brownish-black. **Spores** white (A),
7.5–11 x 6.5–9μm with very low
warts in a fine network. The taste
varies from mild to burningly
acrid. Very common under conifers
in much of North America, it is not edible.

Russula albonigra
Fries

Cap 2–4in/5–10cm. This all-white species soon turns black where bruised or with age. It does not pass through a reddish stage as in the previous species. The gills are rather crowded, cream-buff. **Spores** white (A–B), 7–10 x 6–8μm, with low warts and a partial network. The taste is mild with a menthol-like aftertaste. It is quite frequent under pines. Inedible.

Russula brevipes
Peck

Cap 3–6in/7.5–15cm. This all-white species has narrow, decurrent gills sometimes with a faint blue-green tint. **Spores** white to pale cream (A–C), 8–10.6 x 6.5–8.5μm, with warts up to 1.7μm and a good network. The taste is mild to very peppery in the greenish variety. Common under pines throughout North America. Inedible.

Russula variata
Banning & Peck

Cap 2–6in/5–15cm. This varies a great deal in color, from solid green to completely lavender or mixtures of both. The most distinctive feature is the gills which repeatedly fork from the stem out to the cap margin. **Spores** white (A), 7–10 x 5.5–8μm, with isolated warts. The taste is peppery and the cap cuticle peels about half-way to the center. It is edible and extremely common in woods throughout America.

Russula virescens
(Sch.) Fr.

Cap 2–4in/5–10cm. This attractive species varies from pale blue-green to pale yellow-green or ochre-brown but the cap surface is always cracked into small wooly patches. **Spores** pure white (A), 6–9 x 5.5–7µm with isolated warts or sometimes a few connecting lines. The taste is mild. An uncommon species, it occurs under mixed woods in eastern North America west to Texas. A good edible. The much more common *R. crustosa* also with a green, blue-green, to pale buff, cracking cap differs in its cream spores (C–D).

Russula aeruginea
Lindblad

Cap 2–4in/5–10cm. Varying from pure green to yellowish or grayish-green, often with slightly rusty spotting. **Spores** deep cream (D–E), 6–10 x 5–7µm with low warts and few connecting lines. The taste is mild. A common species in mixed woods throughout North America. Edible.

Russula rugulosa
Peck

Cap 2–4in/5–10cm. The bright red cap is oddly wrinkled and pimpled all over, often shiny and sticky when wet and peels about half-way to the cap center. The gills are pale cream and the stem white. The taste is mild to slightly peppery with an odd, metallic aftertaste. **Spores** deep cream (C–D), 7–9 x 5.5–7 µm, with warts sometimes joined together in short ridges. Although common in mixed woods throughout north-eastern states, it is usually confused with the much rarer *R. emetica*, a species of swampy conifer woods. It has a very hot taste and pure white spores.

Russula betularum
Hora

Cap 1–2in/2.5–5cm. This pretty little species has a delicate pale rose to almost whitish cap, snow-white gills and stem but a fiercely hot taste. **Spores** white (A), 8–10.5 x 6.5–8µm with warts connected by a partial network. Inedible. It is not uncommon in damp woodlands especially under birch.

Russula silvicola
Shaffer

Cap 1–3in/2.5–8cm. Usually a clear pinkish-red, it may also wash out to yellowish-white. The gills and stem are pure white. The taste is painfully hot. **Spores** white (A), 6–10 x 5–9µm, with a well-developed network. It is a common species, often attached to rotten logs or in leaf-litter throughout much of North America.

Russula puellaris
Fries

Cap 1–2in/2.5–5cm. A small, fragile species, the dull reddish-brown to purplish-brown cap discolors dull yellow as do the gills and stem within about 2 hours. The taste is mild and the cap peels about two-thirds to the center. **Spores** pale ochre (D–E), 6.5–9 x 5.5–7µm, with tall, isolated warts. A common species in mixed woods throughout much of America. Edibility is unknown.

Russula polyphylla
Peck

Cap 3–6in/7.5–15cm. This large white, funnel-shaped species soon ages dull pinkish-brown with small scales at the cap center. The decurrent gills are very crowded and pale cream. All parts of the mushroom stain deep reddish-brown when scratched. **Spores** white (A), 7–10 x 5.5–7.5µm. The taste is odd, unpleasant and alkaline, as is the odor. A rather rare species under mixed trees in eastern states. Its edibility is unknown and is best avoided.

Russula claroflava
Grove

Cap 2–4in/5–10cm. The vivid yellow cap, cream gills and slowly blackening flesh characterize this species. **Spores** pale ochre (E–F), 7.5–10 x 6–7.5µm, with more or less isolated warts. The taste is mild and it is frequent in wet, swampy areas under birch. A good edible species.

Russula fucosa
Burlingham

Cap 1–2in/2.5–5cm. The blood-red cap usually has a white bloom and the white stem soon stains brown where handled. The gills are cream. The taste is mild but the odor becomes fish-like with age. If iron-sulphate (FeSO4) is rubbed on the stem it gives a deep olive-green reaction. **Spores** pale ochre (E–F), 8–10.5 x 6–8µm, with tall, isolated warts. A common species in north-eastern states under mixed woods. Edible.

Russula ballouii
Peck

Cap 1–3in/2.5–8cm. The yellow-ochre cap is broken into fine scales especially at the margin. The white stem has scaly yellow patches at the base. The taste is quite peppery. An uncommon but easily recognized species in deciduous woodlands. **Spores** white (A–B), 7–9 x 5.5–7.5μm, with warts joined by a partial network. The very similar *R. burlinghamae* differs in its mild taste and cream spores (C–D).

Russula olivacea
(Schaeffer) Fries

Cap 4–8in/10–20cm. This magnificent species usually has a purplish cap, despite the specific name, but will occasionally be flushed olive. The gills are deep ochre and the white stem is flushed pink, especially at the apex. **Spores** deep ochre (G–H), 8–11 x 7–9μm, with very tall, isolated warts. The taste is mild. A frequent species in some areas under beech.

Russula fragilis
Fries

Cap ½–2in/1.5–5cm. The cap is variable in color usually in mixtures of purple and green, and very fragile. The gills and stem are white. The taste is very peppery. A frequent species in wet areas in mixed woods. **Spores** white (A–B), 6–9 x 5–8μm, with a well-developed network. The similar *R. aquosa* is larger, clearer purple-red with a wet, greasy looking cap and waterlogged stem. The taste is less acrid.

Russula compacta
Frost

Cap 2–6in/5–15cm. The whitish cap and stem soon bruise or discolor reddish-brown and when old develops an odor of fish or crab. Iron-sulphate ($FeSO_4$) on the flesh turns blackish-green. **Spores** white (A–B), 7–10 x 6–8µm, with low warts in a network. A very common species in northern and eastern America in mixed woods. Edible.

Russula vinacea
Burlingham = *R. krombholtzii* Shaffer

Cap 2–4in/5–10cm. The rich purple-red cap, cream gills which are often spotted rust-red, and the often very peppery taste help distinguish this species. **Spores** white (A), 7–9.5 x 6–8µm with a fine partial network. Extremely common in mixed woods throughout the north and east.

Russula earlei
Peck

Cap 2–4in/5–10cm. This remarkable and very primitive species has a thick, waxy cap and very thick, widely spaced gills. The cap surface is often pitted and irregular in outline. The taste is mild to slightly peppery or bitter. **Spores** white (A–B), 5.5–7 x 3.5–5.5µm, with very tiny, isolated warts. This little-known species has proved to be quite common under beech in eastern America. Edibility is unknown.

Russula brunneola
Burlingham
Cap 2–5in/5–12.5cm. The deep brown to purplish-brown cap peels about half-way to the center, while the cream gills are crowded and strongly forked near the stem. **Spores** white (A–B), 6–9.5 x 4–7μm, with low, isolated warts. The taste is mild. A common species in mixed woods. Edible.

Russula lutea
(Hudson) Fries
Cap 2–3in/5–7.5cm. The delicately colored rosy-peach cap contrasts with the deep ochre gills and white stem. The cap cuticle peels completely and the taste is mild. **Spores** deep ochre (G–H), 7–9 x 6–8μm, with rather tall, isolated warts. A rather uncommon species, it is found under deciduous trees in eastern states. Edible.

Russula mariae
Peck
Cap 1–3in/2.5–8cm. One of the most beautiful of mushrooms, the velvety, purple-red cap, stem flushed with pink, and pale ochre gills are distinctive. The cap color varies to yellowish or green. **Spores** pale ochre (D–E), 7.5–9.5 x 6.5–8μm, with a prominent network. The taste is mild to slightly peppery. A very common species under mixed trees, from Canada down to Florida. Edible.

Russula flavida
Frost

Cap 2–4in/5–10cm. The vivid
yellow coloration of cap and
stem are quite unmistakable,
the gills are pale yellow and
the taste is mild. **Spores** cream
(D–E), 5.5–8.5 x 5–7μm, with
warts in a partial network. A
rather uncommon species of
deciduous woods it is edible.

Russula fragrantissima
Romagnesi

Cap 2–6in/5–15cm. The yellow-
ochre cap can be glutinous in wet
weather with a strongly grooved
margin. The odor however is its
most distinctive feature, a strong
blend of marzipan with a sour,
fetid undertone. The taste is oily-
acrid. Very common under mixed
woods throughout America.
Spores cream (C–D), 6.3–9 x
5.5–7.7μm, with a partial network
with large ridges. The even more
fragrant, marzipan-smelling *R.
laurocerasi* differs in its clearer,
paler cap, taller stem and spores
with enormous ridges and flanges.

Russula foetentula
Peck

Cap 2–4in/5–10cm. The orange-
brown cap is coarsely grooved at the
margin while the pale stem becomes
stained bright rust-red from the base
up. The odor is pleasant, rather
marzipan-like while the taste is
acrid-oily. **Spores** pale yellow
(D–E)7–9 x 5.5–7μm, with tall warts
usually isolated. A common species
under mixed woods. Often called
R. subfoetens but that appears to
be a strictly European species. Inedible.

Russula amoenolens
Romagnesi

Cap 2–4in/5–10cm. The rather dull, sepia-brown cap with strong marginal grooves, the fetid, cheesy odor combined with an unpleasant, hot taste are all distinctive. **Spores** cream (B–D), 6–9 x 5–7µm, with mostly isolated warts. It grows commonly under deciduous trees. Inedible.

Russula ventricosipes
Peck

Cap 2–6in/5–15cm. A remarkable species confined to sandy soils under pines in the warmer, eastern and southern states. It is often almost buried in the soil. The cap is dull yellow-ochre and strongly grooved at the margin, while the stem is finely spotted red. **Spores** deep cream (D–E), 7–10 x 4.5–6µm (unusually long) with very minute, isolated warts. The taste is strongly acrid. Inedible.

Russula simillima
Peck

Cap 2–3in/2.5–8cm. The entire mushroom is a dull tawny-yellow to honey color, usually odorless but sometimes smells faintly of household geraniums (Pelargonium). **Spores** white (A–B), 6.5–9 x 5.5–7µm, with a partial network. The taste is peppery. It is frequent under beech trees and is widely distributed. Edibility unknown.

Lactarius deliciosus
(Fr.) S.F. Gray

Cap 2–4in/5–10cm. The orange, zonate cap stains green with age as do the gills and stem. The stem is usually pitted with small spots. When cut the flesh oozes bright orange latex. **Spores** cream 7–10 x 6–7µm, with a fine network. The taste is mild. Often abundant under pines, it is an edible species.

Lactarius thyinos
Smith

Cap 2–4in/5–10cm. This species is uniformly bright orange, including the latex which is unchanging. The stem is usually sticky, especially at the apex. **Spores** cream, 9–12 x 7.5–9µm, with a partial network. It is common in swampy areas under Northern White cedar. Edible.

Lactarius subpurpureus
Peck

Cap 2–4in/5–10cm. The smooth, zonate cap is wine-red to pinkish, mottled with green. The gills are wine-red. The stem is smooth and slimy when wet, wine-red, spotted and pitted with darker red. The latex is scarce, wine-red and tastes slightly peppery. **Spores** cream, 8–11 x 6.5–8µm, ridged and warted with a partial network. Under pines from Maine down through eastern states and common in the south-east. Edible.

Lactarius indigo
(Schw.) Fr.

Cap 2–6in/5–15cm. One of the
world's most remarkable fungi,
the beautiful indigo blue
coloration is quite
unmistakable. The latex is dark
blue turning slowly greenish.
The whole fungus fades to pale
silvery-blue with age. **Spores**
cream, 7–9 x 5.5–7.5μm, with
an almost complete network. A
locally common species under
mixed oak/pine woods in
eastern North America. Edible.

Lactarius torminosus
(Fr.) S.F. Gray

Cap 2–4in/5–10cm. The pale
pink cap is zonate with a very
hairy margin. The latex is
white, unchanging and
painfully acrid. This species is
usually regarded as poisonous.
It occurs under birch in more
northerly states. **Spores**
cream, 7.5–10 x 6–7.5μm with
a network. The similar *L.
pubescens* is paler, less zonate
and less shaggy, and is found
in similar areas.

Lactarius aquifluus
Peck

Cap 2–6in/5–15cm. The cap is
dry, slightly roughened and pale
cinnamon-brown. The latex is
watery, mild. The odor is
distinctive, variously described as
chicory, burnt-sugar or curry.
Spores buff, 6–9 x 5.5–7.5μm,
with mostly isolated warts. A
common species in mixed woods
in northern states. Edibility is
doubtful and best avoided.

Lactarius rufus
(Fr.) Fr.

Cap 2–4in/5–10cm. The dry, bay-red to brick-red cap usually has a central knob. The latex is white, abundant, and slowly very acrid. The brownish stem has a white base. **Spores** cream, 7.5–10.5 x 5–7.5µm with a network. A common species in boggy areas under pine across the country, it is usually regarded as inedible.

Lactarius hygrophoroides
Berk. & Curt.

Cap 2–4in/5–10cm. The cap is dry and rather wrinkled, a bright orange-brown. The gills are very widely spaced and pale cream. The latex is white, very abundant and mild to taste. **Spores** white, 7.5–10.5 x 6–7.5µm with a network. This species is very common under deciduous trees in eastern and northern North America. The closely related *L. volemus* differs in its crowded gills (see illustration) and the brown stains that develop on the broken gills.

Lactarius corrugis
Peck

Cap 2–6in/5–15cm. The rich liver-brown to reddish-brown cap is conspicuously wrinkled-corrugated at the margin and the gills are deep ochre. The latex is white, copious and mild. The odor can be fish-like when old. **Spores** white, 9–12 x 8.5–12µm with a network. A good edible species, it is frequent under deciduous trees in eastern and northern North America.

Lactarius lignyotus

Fries

Cap 1–4in/2.5–10cm. The blackish-brown cap is velvety and slightly wrinkled contrasting with the almost white gills. The latex is white, abundant, and soon discolors the damaged tissues bright-rose. The taste is slightly peppery. **Spores** bright ochre, 9–10.5 x 9–10µm, with a network. It is a common species under conifers in mossy areas in eastern and northern states.

Lactarius camphoratus

(Fr.) Fr.

Cap 1–2in/2.5–5cm. The deep reddish-brown to liver-colored cap has a small knob at the center while the gills and stem are cinnamon to ochre-brown. The latex is watery white, scanty and mild. As it dries the mushroom gives off a powerful scent of curry, burnt sugar or chicory. **Spores** yellowish, 7–8.5 x 6–7.5µm, with a few connecting lines. Very common in mixed woods in eastern and northern states.

Lactarius deceptivus

Peck

Cap 2–10in/5–25cm. This often very large species has a white, depressed cap with a strongly inrolled, wooly margin at first (best seen by cutting in half). The white, crowded gills are slightly decurrent. The latex is white, strongly acrid. The stem is very short, stout, firm and velvety. **Spores** pale buff, 9–13 x 7.5–9µm, with faint connecting lines. A very common species in mixed woods in eastern woods west to Michigan and Ohio. Inedible.

Lactarius piperatus
(Fr.) S.F. Gray

Cap 2–6in/5–15cm. The cap is soon funnel-shaped, dry and slightly velvety. The gills are extremely crowded, white to cream. The latex is white and very acrid. In some forms it dries olive-green. **Spores** white, 6–8 x 5–5.5μm, with a faint network. This inedible species is frequent under deciduous trees in eastern North America west to Michigan.

Lactarius thejogalus
(Fr.) S.F. Gray

Cap 1–3in/2–8cm. The cap often has a sharp umbo and is a brick-red to cinnamon-orange color, fading with age. The gills and stem are concolorous or paler. The latex is white then slowly turns yellow, and is rather acrid to taste. **Spores** cream, 7–9 x 6–7.5μm, with tall ridges forming a partial network. This is a common species in wet, boggy habitats in mixed woods over much of eastern and northern United States. Edibility is doubtful and it is best avoided.

Lactarius chrysorheus
Fries

Cap 1–3in/2–8cm. The pale, yellowish-cream to slightly pinkish cap is often marked with zones or watery spots. The gills and stem are pale cream to buff. The latex is white but turns rapidly bright yellow as does the flesh when cut. The taste is slowly acrid. **Spores** pale yellow, 6–8 x 5.5–6.5μm, with isolated warts and a slight network. Common under oaks in much of North America. One should also look out for the very similar *R. vinaceorufescens,* which differs in developing dull vinaceous-brown stains over most of the mushroom. Neither species are good edibles.

Lactarius glyciosmus
(Fr.) Fries

Cap 1–3in/3–8cm. The entire mushroom is pale gray with a hint of pink or lavender but the most distinctive character is the pleasant odor of dried coconut. The latex is white, unchanging and tastes very slightly acrid. **Spores** cream, 6–8 x 5–6µm, with bands and ridges forming a broken network. A common species, it is edible but of poor quality.

Lactarius atroviridis
Peck

Cap 2–6in/5–15cm. A remarkable species and quite unmistakable with its deep, blackish-green cap and stem. The latex is white and very acrid. **Spores** cream, 7–9 x 5.5–6.5µm, with a partial network. A rare species under mixed trees in eastern North America, it can be very difficult to see against the fallen leaves. Inedible.

Lactarius quietus
(Fr.) Fries

Cap 2–4in/5–10cm. The dull, reddish-brown cap is usually zoned with darker bands and the gills and stem are a similar color. The latex is white to slightly yellowish, mild then soon slightly acrid. The odor is distinctive but hard to describe, most people say it is oily-sweet. **Spores** cream, 7–9 x 5.5–7µm, with mostly isolated warts and ridges. It has been found quite frequently in north-eastern states under oaks. It is edible but poor.

HYGROPHORUS FAMILY

All these mushrooms share the common character of thick, waxy gills, and smooth white spores formed on very long basidia. *Hygrophorus* species are thicker, fleshier and usually duller colored than the fragile, often brilliantly colored *Hygrocybe*.

Hygrocybe pratensis
Persoon.
Cap 2–4in/5–10cm. The cap soon becomes flattened, often with a central bump and the gills run down the stem. The whole fungus is pale orange-buff. **Spores** white, 6–7 x 4–5µm. A common species it occurs in grassy areas in woods and fields. Edible and quite good.

Hygrocybe virginea
(Wulf. ex Fr.) Orton & Watling (= *H. niveus*)
Cap 1–2in/2.5–5cm. The entire mushroom is pure white, smooth to slightly greasy, hygrophanous, with a translucent, striate cap margin when wet. The gills are distant and run down the stem (decurrent). **Spores** white, 7–10 x 5–6µm. A common species in grassy, woodland clearings and fields in western states. In the east it is usually replaced by the almost identical *H. borealis* which has a dry cap. Edible but worthless.

Hygrocybe conica
(Scop.) Kummer.
Cap 1–2in/2.5–5cm. The often acutely conical cap is bright red while the stem is yellowish. The gills are pallid to yellow. All parts of the fungus stain black on handling and after a few hours the entire mushroom will be blackened. **Spores** white, 8–12 x 5–7µm. A common species in woodlands and grassy clearings throughout America.

Hygrocybe punicea
(Fr.) Kummer

Cap 2–4in/5–10cm. This large, beautiful species is a bright scarlet to blood-red with yellowish-orange gills. The base of the stem and the flesh inside are white. **Spores** white, 8–11 x 5–6μm. Widespread in woodlands and grassy areas especially in western states late in the year through January. Edibility has been questioned, possibly poisonous.

Hygrocybe miniata
(Fr.) Kummer

Cap 1–2in/2.5–5cm. The brilliant orange-red to scarlet cap is minutely scurfy-scaly at the center. The gills are orange-red and broadly attached to slightly decurrent. **Spores** white, 7.5–11 x 5–6μm. An often very common species it is found in damp woodlands, often in moss on rotting logs and stumps. The very similar and equally common *H. quieta* (Kuhn.) Singer differs in its truly decurrent gills and the unpleasant oily smell it leaves on fingers when the fungus is crushed and rubbed. Edible but bland.

Hygrocybe psittacina
(Schff. ex Fr.) Wuensche

Cap ½–1in/1–2.5cm. This remarkable species is often called the Parrot because its cap varies from intense green to blue-green, yellow to orange and can change within 2 hours. The surface is always very glutinous on both cap and stem. The gills are greenish to yellow and adnate. **Spores** white, 8–10 x 4–5μm. A common species in grassy clearings and woodlands across America. Edible but worthless.

Hygrocybe laeta

(Pers. ex Fr.) Karsten
Cap ½–1in/1–3cm. The slimy-glutinous cap varies from orange-brown to pinkish-brown while the stem often has a greenish or grayish apex and is also slimy. The gills are decurrent and grayish-pink. **Spores** white, 6–8 x 4–5μm. A widely distributed species in woods and clearings across America. Edible but worthless.

Hygrocybe unguinosa

(Fr.) Karsten
Cap 1–2in/2.5–5cm. Unmistakable with its entirely gray, very viscid-glutinous cap and stem. The gills are thick, deeply adnate and also gray. **Spores** white, 6–8 x 4–6μm. An uncommon species but widely distributed throughout much of America. Edible but worthless.

Hygrocybe ovina

(Bull. ex Fr.) Kuehn.
Cap 1–3in/2.5–8cm. This somber, dark gray-black to brownish-black mushroom bruises reddish wherever it is handled or broken and often has an odor of ammonia or bleach. **Spores** white, 7–12 x 4.5–6μm. An uncommon species it is difficult to spot against the dark leaf-litter in which it likes to grow, in deciduous woods throughout eastern states. Edibility doubtful.

Hygrophorus chrysodon
(Batsch) Fr.

Cap 1–3in/3–8cm. This all-white
fungus is distinctive for its stem apex
and cap margin are speckled with
bright golden-yellow flecks. The white
gills run down the stem. The smell is
faint, said to resemble Jerusalem
artichokes. **Spores** white, 8–10 x
4–5μm. An uncommon species, found
under pines and oak especially in
western states. Edible but poor.

Hygrophorus penarius
Fries (= *H. sordidus* Peck)

Cap 4–6in/10–15cm. One of the largest
of the white *Hygrophorus*
species, the thick, creamy gills
are widely spaced and run
down the stem a little. The
cap is dry to slightly downy to touch.
Spores white, 6–8 x 4–5μm. Found
under beech trees, it is not uncommon
in eastern and northern states. Often
called *H. sordidus* Peck but that appears
to be identical with this species which
was described much earlier. Edible.

Hygrophorus russula
(Sch.) Quel.

Cap 3–6in/8–15cm. The specific
name refers to its resemblance to
mushrooms of the genus *Russula,* and
it does indeed look like one with its
reddish cap and short, squat stature.
The cap is usually speckled with
darker, reddish-brown to wine-red
spots on a paler background. The
rather crowded gills are pinkish-
cream, also often spotted wine-red.
Spores are white, 6–8 x 4–6μm.
Often growing in large groups under
beech and oak late in the year in the
north-east and across to the northwest. Edible.

Hygrophorus hypothejus
(Fr.) Fries

Cap 2–4in/5–10cm. This very
glutinous, viscid species has a
yellowish-brown to olive-brown cap,
yellowish gills and stem with a slight
annular zone at the stem apex.
Spores white, 8–9 x 4–5µm. A
common species under conifers
particularly at the end of the year,
often after the first frosts. Widely
distributed across the United States.
Edible but poor quality.

PLEUROTUS FAMILY

These genera and species all grow on wood and usually
have the stem reduced or completely absent. The spores
vary from white to pale lavender to faintly pinkish.

Pleurotus ostreatus   Oyster Caps
(Jacq.) Kummer.

Cap 2–6in/5–15cm. The caps are semicircular and broadly
attached at the rear to the wood from which they grow. They
frequently grow in large numbers, forming overlapping
clusters. The color varies from deep bluish-black to pale gray-
brown. The gills are cream, narrow and very crowded. **Spores**
pale lilac, 7–11 x 3–4µm. Found on dead or dying trees,
usually late in the year, often even in frosty weather. The
equally common *P. pulmonarius*
occurs much earlier
in the season and is
much paler, often
almost white.
Common and a
good edible species.

Pleurotus cornucopiae

Paul ex Fr.(= *P. sapidus*)
Cap 2–6in/5–15cm. This species is
unusual in having an almost
central stem and forming a more
funnel-shaped structure. The
overall color is pale buff to cream.
The gills are deeply decurrent and
tend to form a mesh at the base.
Spores lilac, 8–11 x 3.5–5µm.
Frequent on dead deciduous trees
and is widely distributed. Edible
and good.

Phyllotopsis nidulans

(Pers. ex Fr.) Singer
Cap 1–3in/2.5–8cm. The bright
yellow-orange cap is semicircular to
kidney-shaped and hairy. The gills are
crowded and a deeper shade of
yellow-orange. The odor is strong and
unpleasant, almost fetid. **Spores** pale
yellowish pink, 4–5 x 2–3µm. A
common fungus on dead wood
throughout much of North America.
In eastern states, also look for *P.
subnidulans* which has a darker orange
cap, thinner, more widely spaced gills
and curved, sausage-like spores.
Inedible.

Panus tigrinus

(Bull. ex Fr.) Singer
Cap 1–3in/2.5–8cm. The
domed cap is almost white
with small blackish scales,
while the gills are cream,
minutely serrated and run
down the stem. **Spores**
white, 7–8 x 3–3.5µm.
Not uncommon, this
species grows on dead
deciduous trees, especially
in northern states. Edible
but tough.

Panellus serotinus

(Hoffm.) Kuhn.
Cap 2–4in/5–10cm. The cap
is broadly attached at the
rear or sometimes has a very
short stem, the color varies
from deep olive-green to
brownish-yellow, often with
a violet flush and the surface
is slightly velvety. The gills
are pale yellow-orange and
crowded. **Spores**
yellowish, 4–6 x 1–2μm.
Fairly common on dead
timber from the Pacific
Northwest across to
eastern states, usually in the
colder weather at the
end of the season.
Edible but not
recommended.

Panellus stipticus

(Bull.) Karsten
Cap ½–1in/1–2.5cm. The shell-
shaped cap has a small but distinct
stem at one edge. The surface of the
cap is dull ochre-buff and minutely
velvety. The gills are crowded, pale
buff and often forked. **Spores** white,
3–5 x 1.5–3μm. A very common
fungus on fallen deciduous wood,
especially in the east. Its most
remarkable feature is that the gills
glow bright greenish-yellow in the
dark. Inedible, the taste is bitter or
acrid.

TRICHOLOMA FAMILY

An enormous family of mushrooms varying greatly in appearance and very difficult to characterize. They all share white to pale pinkish spores. Some (*Clitocybe*) have decurrent gills, others (*Tricholoma*) have sinuate gills. They range in size from tiny to huge, and some genera have veils.

Omphalina pyxidata

(Bull.) Quel.

Cap ¼–1in/0.4–2.5cm. The funnel-shaped cap is rusty-brown with a fluted margin. The decurrent gills are slightly paler as is the rather short stem. **Spores** white, 7–10 x 4.5–6μm. A common species in grassy woodlands and lawns everywhere. Edibility unknown.

Rickenella fibula

(Bull. ex Fr.) Raith

Cap ¼–½in/0.4–0.6cm. Depressed at center and fluted on margin, bright orange-brown to reddish-brown. Gills are widely spaced and decurrent. The stem is slender and rather tall in proportion to the cap. **Spores** white, 4–5 x 2–2.5μm. A common species in mossy areas in woods and fields, widely distributed.

Cantharellula umbonata

(Gmel. ex Fr.) Singer

Cap 1–2in/2.5–5cm. Soon depressed at the center with a central umbo, the cap is gray to blackish-gray and often has a violet flush. The gills are paler, cream, rather thick and strongly forking. The stem is tall and colored like the cap. The flesh when damaged turns reddish-brown. **Spores** white, 7–11 x 3–4μm. A frequent species in wet pine woods in mossy areas throughout the north-east. Edibility doubtful, best avoided.

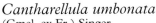

Xeromphalina campanella

(Batsch. ex Fr.) R. Maire
Cap ¼–½in/0.4–1cm. This
species grows in enormous
numbers forming beautiful
drifts over decaying conifer
stumps and logs. The caps are
bright golden orange while
the decurrent gills are yellow.
Spores white, 5–8 x 3–4µm.
Widespread throughout
North America. *X. kauffmanii*
is similar but grows in the east
on hardwoods. Too small to
be edible.

Armillaria mellea Honey Fungus

(Vahl.) Kummer
Cap 2–4in/5–10cm. The
honey-yellow to greenish cap
is almost smooth or may
have very fine, brownish
scales at the center. The gills
are adnate to slightly
decurrent, creamy-white to
pinkish. The tough, fibrous
stem is swollen, pale honey-
yellow to greenish-buff and
often with a yellow, wooly
coating at the tapered base.
There is a well-developed
wooly ring above, white to
yellow. **Spores** white, 7–10
x 5–6µm, smooth. Often in
huge clumps on dead or
dying trees, if the bark is
peeled back then long, very
tough, black "bootlaces"
may be found. These
rhizomorphs are the Honey
Mushroom's way of
spreading long distances to
infect other trees. It is a
deadly parasite of forest
trees. Edible and good when
cooked well.

Armillaria bulbosa Honey Fungus
(Barla) Romagn.

Cap 2–4in/5–10cm. The pale
reddish-brown to pinkish-brown caps
have darker brown scales at the center
and remains of white veil at the
margin. The gills are cream to pale
pinkish-buff. The swollen, club-
shaped stems are pale pinkish-buff
with a white, cobwebby veil above
and often a yellowish coating at the
base. **Spores** white, 7.5–8.5 x
4.5–5µm, smooth. Often in large
numbers but may be scattered over a
large area, usually on the ground on
buried wood or roots of dying trees.
Edible when well cooked.

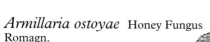

Armillaria ostoyae Honey Fungus
Romagn.

Cap 2–4in/5–10cm. The caps are
pale to dark reddish-brown with dark,
blackish-brown scales. The gills are
cream to pale pinkish-buff. The stems
are tapered, pale brown with a white,
wooly ring that has dark brown scales
on the edge and undersurface. **Spores**
white, 8–10 x 5–6µm, smooth.
Common, in large clumps on dead
or dying timber, widely distributed.
Edible when well cooked.

Armillaria tabescens
(Scop. ex Fr.) Emel.

Cap 2–3in/5–7.5cm. The ochre-
brown to honey-yellow caps are
smooth with tiny brown scales at the
center. The gills are adnate to slightly
decurrent and pale pinkish-buff. The
slender stem is completely free of any
ring or veil. **Spores** white, 8–10 x
5–7µm. Often growing in enormous
clusters at the base of oak and other
deciduous trees, in summer and early
fall. Edible.

Hygrophoropsis aurantiaca False Chanterelle
Maire (Wulf. ex Fr.)

Cap 1–3in/2.5–7.5cm. This species is white-spored but has a similarly soft, inrolled, cap margin like *Paxillus*. Color ranges from pale yellow to orange while the decurrent, forked and rather blunt gills are bright orange. **Spores** 5–8 x 3–4.5μm. It grows under pine and birch in damp locations throughout North America. It is sometimes picked in mistake for the edible Chanterelle. It is not poisonous, just of poor quality.

Clitocybe clavipes
(Pers. ex Fr.) Kummer

Cap 2–3in/5–7.5cm. The dull gray-brown cap contrasts with the pale yellowish-cream, deeply decurrent gills. The grayish stem is swollen, club-shaped at the base. This mushroom often has a sweet, fragrant odor. **Spores** white, 6–8.5 x 3–5μm. A common species under pines especially in the east but also occurs in western states. Inedible, reported to cause poisoning when consumed with alcohol.

Clitocybe geotropa
(Bull. ex Fr.) Quel.

Cap 2–8in/5–20cm. This often very large species soon becomes funnel-shaped with a central umbo. The color ranges from dull ivory-white to ochre and the cap surface is finely roughened. The decurrent gills are pale cream and crowded. **Spores** are white, 6–7 x 5–6μm. An uncommon species, it is found in mixed woodlands, particularly in the east, although it extends across to western states. Edibility doubtful and best avoided; other white *Clitocybe* species are poisonous.

Clitocybe gibba

(Pers. ex Fr.) Kummer = *C. infundibuliformis*

Cap 2–4in/5–10cm. The cap is a delicate pinkish-buff to tan and smooth, while the decurrent gills are whitish, as is the stem. **Spores** white, 5–8 x 3.5–5µm. This rather delicate, thin-fleshed mushroom is very common in mixed woods, especially deciduous, throughout America. The similar *C. squamulosa* has a slightly roughened, darker cap with a concolorous stem and occurs under conifers, especially in the spring and early summer. Edible but not recommended.

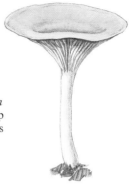

Clitocybe nebularis

(Batsch) Kummer

Cap 2–6in/5–15cm. The overall smoky-brown colors, with crowded but hardly decurrent gills, are distinctive features along with the late occurrence in the season in piles of leaf-litter. **Spores** pale buff, 6–7 x 3–4µm. It often grows in fairy rings and is quite common on the west coast, usually from December on. Edible but poor, not recommended.

Clitocybe odora

(Bull. ex Fr.) Kummer

Cap 1–3in/2.5–8cm. The delicate blue-green to gray-green colors, lovely odor of anise, smooth cap and pinkish-buff spores all characterize this beautiful mushroom. **Spores** 6–7 x 3–4µm. Widely distributed throughout America in mixed woods. In eastern states the equally common *C. aeruginosa* is also blue-green but lacks the odor, has a minutely hairy cap, and has white spores. Both are edible.

Clitocybe rivulosa
(Pers. ex Fr.) Kummer
Cap 1–2in/2.5–5cm. The whitish cap has a "frosty" coating which is usually cracked in concentric zones showing a browner undersurface. The crowded gills are white and adnate. **Spores** white, 4–5.5 x 2.5–3μm. A frequent species in grassy areas late in the season and widely distributed. An extremely poisonous species which should be avoided, as should all small, white mushrooms.

Clitocybe nuda
(Bull.ex Fr.) Big. & A. H. Smith
Cap 2–6in/5–15cm. The lovely, smooth, violet cap soon fades to a violet-tan but the crowded gills remain a pale lavender-violet. The stem is tough, fibrillose and pale violet. The odor is often rather fragrant. **Spores** pale pinkish-buff, 6–8 x 4–5μm and minutely roughened. A common species, often in large numbers in circles on beds of leaf-litter or compost throughout America. Edible and very popular.

Clitocybe inversa
(Scop. ex Fr.)
Cap 2–4in/5–10cm. The funnel-shaped cap is a rich tawny-orange to foxy-red. The decurrent gills are a paler orange. **Spores** pale creamy-yellow, 4–5 x 3–4μm, minutely roughened. It often grows in groups in mixed woods especially in western states. Edibility doubtful, not recommended.

Laccaria laccata

(Scop. ex Fr.) Bk. & Br.

Cap ½–2in/1–5cm. One of the most confusing mushrooms because it has many variations in color, size and shape. However, if the characters of thick, pinkish gills, pinkish-brown to reddish-brown cap and fibrous stem, and white spore–print are observed then identification is not too difficult. **Spores** 7–9 x 6–8μm and minutely spiny. One of the most common and widespread species over the northern hemisphere, found in mixed woods, bogs and open moorland with trees. Edible.

Laccaria nobilis

Mueller

Cap 2–4in/5–10cm. This is one of the largest species of *Laccaria* known, even though it was only recently described. The reddish-brown cap and stem are both rather scaly-fibrous while the gills are thick and pale pink. The odor is often rather radishy. **Spores** white, 7.5–10 x 6.5–8.5μm. Not uncommon in eastern and northern states where it occurs in damp, boggy woods especially under pines. Edible.

Laccaria amethystea
(Bolt. ex Hooker) Murrill

Cap 1–2in/2.5–5cm. The beautiful deep amethyst-violet of the entire fungus when fresh and moist is quite unmistakable, but as it dries the cap becomes a dull, grayish-lavender to almost white. The gills however remain deep violet. **Spores** white, globose, 8–10µm, minutely spiny. Common in shady, damp woodlands, both deciduous and conifer, especially in the east and north. In western states it is replaced by the very similar *L. amethysteo-occidentalis* which has elliptical spores. Both are edible.

Laccaria ochropurpurea
(Berk.) Peck

Cap 2–8in/5–20cm. The largest and most robust of the species *Laccaria*, the pale grayish-brown to slightly purplish cap and stem contrast with the thick, deep purple gills. Both cap and stem may be rather scaly-fibrous. **Spores** white, globose, 6–8µm, minutely spiny. This species shows a preference for drier, mixed woods, especially oak in eastern states. Edible. In dry, coastal areas *L. trullisata* occurs, with a large, swollen stem often buried deep in sand.

Laccaria tortilis
(Bolt.) S. F. Gray

¼–½in/½–1cm. Perhaps the smallest species of *Laccaria*, distinguished by the thick gills which are few in number, and the wavy–fluted cap which is pale pinkish-brown. **Spores** white, globose, 11–16µm and spiny, and with only 2 spores per basidium. Rather uncommon, in very damp areas, often on bare soil in deep shade. Edible but worthless.

Tricholoma sulphureum
(Bull. ex Fr.) Kummer
Cap 2–4in/5–10cm. The overall
sulfur-yellow coloration is
combined with a strong and
pungent odor of coal-gas, making
an easily recognized combination.
Spores white, 9–12 x 5–6µm.
Fairly common in mixed woods in
the Pacific Northwest. Inedible,
possibly poisonous. In eastern
states the very similar *T. odorum*
has the gas odor and yellow colors
but differs in its pallid, whitish
gills.

Tricholoma saponaceum
(Fr.) Kummer
Cap 2–4in/5–10cm. This is a very
variable mushroom with colors
ranging from gray-brown to greenish-
brown and often with pink flushes in
the stem base. The cap is smooth to
minutely scaly. The odor is rather
fragrant, soapy. **Spores** white, 5–6 x
3.5–4µm. Often very common,
especially in the Pacific Northwest,
but also in eastern states, usually in
mixed woods. Inedible, but may be
poisonous.

Tricholoma sulphurescens
Bresadola
Cap 2–6in/5–15cm. A large,
attractive species with white cap
and stem but with all parts
staining sulfur-yellow when
handled. The odor can be rank
and unpleasant but the taste is
mild. **Spores** white, 5–6 x
4–5µm. This is a rare species
and its distribution is uncertain,
but it has been found under
conifers in the north-
east. Inedible.

Tricholoma terreum
(Sch.) Kummer

Cap 2–4in/5–10cm. The cap is dull gray to brownish-gray, felty-scaly with darker fibers. The gills and stem are white. The odor and taste are mild. **Spores** white, 5–7 x 4–5μm. An often abundant species under pines late in the season, from east to west. Edibility uncertain, best avoided.

Tricholoma virgatum
(Fr.) Kummer

Cap 2–4in/5–10cm. The conical cap is silvery-gray with darker gray-black fibers. The gills are grayish to yellow-gray while the stem is white. The taste is sharp and peppery. **Spores** white, 6–7 x 5–6μm. Found under conifers throughout North America. Inedible.

Tricholoma subresplendens
Peck

Cap 2–4in/5–10cm. All parts of this mushroom are a beautiful snow-white, only slightly brownish when old. **Spores** white, 5–7 x 3.5–4.5μm. Quite common under oaks in eastern North America. Edibility uncertain, best avoided.

Tricholoma sejunctum
(Sow. ex Fr.) Quel.
Cap 2–4in/5–10cm. The pale greenish-yellow cap is marked with darker, blackish fibers, while the white gills and stem may be flushed with yellow. Both the taste and odor are mealy, cucumber-like, taste becoming bitter with age.
Spores white, 5–6 x 4–5μm. Fairly common under mixed trees in the Pacific Northwest and eastern states.
Inedible.

Tricholoma flavovirens
(Pers. ex Fr.) Lund. (= *T. equestre*)
Cap 2–4in/5–10cm. The yellow cap is marked with dark brownish to olive fibers and is usually dry. The stem and gills are both yellow.
Spores white, 6–8 x 3–5μm. Usually found under pines in sandy soils, it is widely distributed throughout North America.
Edible and good.

Tricholoma portentosum
(Fr.) Quel.
Cap 2–5in/5–12.5cm. The smooth, grayish cap is often flushed with yellow and is marked with darker, blackish fibers. The stout, white stem and white gills are often also flushed with yellow. The odor and taste are mealy, cucumber-like. **Spores** white, 5–6 x 3.5–5μm. A common species under pines late in the season throughout North America.
Edible.

Tricholoma fulvum

(D.C.) Sacc. (= *T. flavobrunneum*)
Cap 2–4in/5–10cm. The
moist, orange-brown to red-
brown cap often has a
grooved margin. The gills are
yellowish spotted with reddish-
brown while the stem is fibrous,
reddish-brown. The flesh in the
stem is yellow. **Spores** white,
5–7 x 3–4.5µm. This species,
common in Europe under
birch, has been reported in
similar habitats in northern
and western states. Inedible.

Tricholoma aurantium

(Sch.) Ricken
Cap 2–4in/5–10cm. This attractive
species is a bright orange to orange-
brown with a viscid cap with darker
brownish flecks. The gills are white,
often spotted orange, while the stem
has concentric orange-scaly zones
over the lower half. Odor and taste are
mealy cucumber-like. **Spores** 4–5 x
3–3.5µm. A frequent species under
conifers late in the season,
throughout much of North America.
Inedible.

Tricholoma caligatum

(Viv.) Ricken (= *Armillaria caligata*)
Cap 2–6in/5–15cm. The dull, pallid
cap has darker brown scales and
fibers. The stem is white with a sheath
of brown veil broken up into patches
and zones over the lower half. There
may be bluish stains present in some
forms. **Spores** white, 6–7.5 x
4.5–5.5µm. The odor is often spicy or
pungent while the taste can be
disagreeable. Under deciduous trees
in the east, under conifers in the west,
quite common. Edible and good.

Leucopaxillus albissimus
(Peck) Singer

Cap 2–4in/5–10cm. This all-white
mushroom is fleshy with crowded,
adnate to slightly descending gills.
The odor is usually fragrant while the
taste is bitter. The base of the stem is
bound to the leaf-litter by copious
white masses of mycelium. **Spores**
white, 5.5–8.5 x 4–6μm, ornamented
with tiny amyloid warts. Often in large
circles in conifer woods in northern
North America. Inedible.

Melanoleuca alboflavida
(Peck) Murrill

Cap 2–6in/5–15cm. This tall,
slender mushroom typically
has a cap with a central bump,
buff-cream colors and dry to
moist surface. The cream gills
are narrow and very crowded.
The stem is tall, brittle and
fibrous. **Spores** white, 7–9 x
4–5.5μm, with amyloid warts.
Special harpoon-like cells
(cystidia) are present on the
gills. A common species in
deciduous woods in eastern
states. Edible.

Melanoleuca cognata
(Fr.) K. & M.

Cap 2–4in/5–10cm.
The domed cap is a rich yellow-
ochre to yellow-brown while the
gills are deep ochre. The fibrous
stem is paler, cream-ochre.
The odor can be floury-rancid.
Spores are unusual for the group,
being deep creamy-yellow, 9–10 x
6–7μm, with amyloid warts. In
coniferous woods, especially spruce
in the Rocky Mountains and
south-west. Edible.

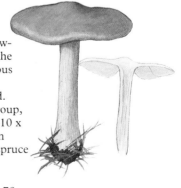

Lyophyllum decastes

(Fr.) Singer (= *L. aggregatum*)
Cap 2–4in/5–10cm. Often growing
in enormous clumps the dull gray-
brown caps are smooth and sometimes
irregular in shape. The gills are white
to buff and attached to the stem,
sometimes with a decurrent tooth.
The tough, fibrous stems are white to
gray. **Spores** white, globose, 4–6µm.
Usually found in disturbed soils on
roadsides, gardens, woodland edges,
very common in some years over most
of America. Edible.

Catathelasma imperialis

(Fr.) Singer
Cap 5–15in/12.5–38cm. This
mushroom has a size to match
the length of its name. The huge caps
are dull brown breaking into broad
patches and scales. The crowded gills
descend down the stem and are buff-
yellow. The thick, squat stem has a
double veil of pinkish-brown ending
in a double ring at the top of the
stem. The white flesh is very thick
and firm. **Spores** white, 10–15 x
4–5.5µm, amyloid. In groups under
spruce and fir in the Rocky Mountains
and Pacific Northwest. Edible but tough.

Calocybe carnea

(Bull. ex Fr.) Kuhn.
Cap ½–2in/1.5–5cm. This small,
pink to reddish-brick mushroom has
smooth, dry cap and stem. The gills
are white, crowded, attached to
slightly descending. The stem is
fibrous in texture. **Spores** white,
4–6 x 2–3µm, smooth. This pretty
species grows in grass in lawns
and open woodland clearings, rather
uncommonly, over much of Northern
North America. Inedible.

Tricholomopsis rutilans
(Schff. ex Fr.) Singer

Cap 2–6in/5–15cm. The contrasting colors of bright wine-red cap and stem, against the yellow gills, and the habitat on conifer wood, allows easy recognition. The cap is minutely velvety-scaly. **Spores** white, 7–8 x 5–6μm. A common species on dead and decaying pine stumps and logs throughout North America. Edible but poor.

Tricholomopsis platyphylla
(Pers. ex Fr.) Singer

Cap 2–6in/5–15cm. One of the earliest mushrooms to appear, the dull gray-brown cap is streaked with radial fibers. The gills are very broad, widely spaced and often with split or jagged edges. The white stem is very tough and fibrous and has white root-like cords at the base (rhizomorphs). **Spores** white, 7–9 x 5–7μm. Common on deciduous logs, stumps and buried wood throughout eastern North America, rarer in the west. Inedible.

Asterophora lycoperdoides
Bull. ex Mer.) Dit.
Cap ¼–½in/0.6–1.2cm. A
remarkable small mushroom found on
the decaying remains of old fungi,
specifically *Russula* and *Lactarius*. The
cap is covered in a thick brown
powdery coating which is actually
masses of asexual spores. The thick
gills are widely spaced and often
malformed. Widely distributed
throughout North America.
Inedible. The related
A. parasitica differs in
forming a perfect small,
white mushroom without
the powdery coating, and
also grows on rotting
Russula and *Lactarius*.

Flammulina velutipes
(Fr.) Kar.
Cap 1–4in/2.5–10cm. Often fruiting
while snow is still on the ground, the
rich yellow-orange caps are slightly
sticky. The gills are crowded, cream
while the stem is yellow with a
blackish-brown, hairy base. **Spores**
white, 7–9 x 3–6μm. Growing in
dense clusters on standing trees and
stumps, especially elm, aspen and
willow, throughout North America,
from October to May. Edible and
good.

Cystoderma terrei
(Bk. & Br.) Harmaja
= *C. cinnabarinum*
Cap 1–3in/2.5–7.5cm. The rich
brick-red to orange cap is very
powdery-granular as is the stem. The
gills and stem apex are white. **Spores**
white, 3.5–5 x 2.5–3μm. Uncommon
in mixed woods throughout America.
Inedible.

Cystoderma amianthinum
(Scop.) Fayod

Cap 1–2in/2.5–5cm. The powdery-granular cap is bright yellow-ochre and usually wrinkled at the center. The stem has a yellow, granular coating up to a slight ring-zone near the apex. The gills are white to cream. **Spores** white, 4–6 x 3–4μm. Common in conifer woods throughout North America, especially in the north. Inedible.

Xerula furfuracea
Redhead = *Oudemansiella radicata* of some authors.

Cap 2–5in/5–12.5cm. This tall, elegant species, if carefully dug out of the soil, will reveal a long, deeply rooting "tap-root." The pale brownish cap is glutinous in wet weather but minutely hairy when dry. The stem is dry and minutely velvety. **Spores** white, 12–18 x 9–12μm. Common around dead stumps throughout much of North America. Inedible. The true *O. radicata* appears to be found only in Europe, and like this species has recently been placed in the genus *Xerula*.

Squamanita odorata
(Sum.) Bas
Cap 1–3in/2.5–7.5cm. This rare mushroom is unusual in having a tuberous underground base to its stem. The cap is conical and fibrous and lilac-brown. The stem is also lilac-brown and fibrous up to a slight ring-zone, while the gills are white. **Spores** white, 6.5–9 x 4–6µm, smooth. The odor is quite strong, aromatic or fruity. Found in sandy or poor soils in woods, especially in the east, but across to Washington also. Edibility unknown.

Marasmius androsaceus
(L. ex Fr.) Fr. Horsehair Fungus
Cap ¼–½in/1–2cm. The small, fluted brown cap on the thin, hair-like, blackish-brown stem are distinctive as are the very widely spaced gills. There is no odor, unlike some other *Marasmius* species. **Spores** white, 6–9 x 2.5–4.5µm. Often abundant on conifer needles and twigs in northern and western states. Inedible.

Marasmius rotula
(Scop. ex Fr.)Fr.
Cap ¹⁄₁₆–½in/0.15–1.5cm. The tiny, bell-shaped cap is white, pleated and sunken at the center. The white gills are widely spaced, attached to the stem or separated by a narrow "collar." The thread-like stem is dark brown and shiny. It has no odor. **Spores** white, 6–10 x 3–5µm. Common on dead wood and twigs of deciduous trees throughout North America. Inedible.

Marasmius oreades Fairy Ring Mushroom
(Bolton) Fr.

Cap 1–3in/2.5–7.5cm. Although other mushrooms also grow in rings this is the one most people are aware of since it disfigures lawns and grassy areas throughout much of the world. The tough, pale buff caps and fibrous stems appear whenever the weather is mild and damp. The gills are thick and widely spaced. **Spores** white, 7–10 x 4–6µm. Edible and delicious although care should be taken not to confuse it with poisonous *Clitocybe* species which also grow in lawns. The latter are usually white, flattened to depressed caps with crowded gills.

Marasmius siccus
(Schw.) Fr.

Cap 1⁄8–1¼in/0.3–3cm. The beautiful deep orange-rust caps are bell-shaped, deeply fluted with a sunken center. The gills are pale cream-yellow. The slender stem is yellowish-brown, smooth and brittle. **Spores** white, 16–21 x 3–4.5µm. Common on twigs and leaves in mixed woods especially in eastern states. Inedible. *M. fulvoferrugineus* is deeper rust-brown.

Marasmius nigrodiscus
(Peck) Halling

Cap 2–4in/5–10cm. The pallid ivory-buff caps have a darker brown disc and are translucent when wet, becoming opaque when dry. The gills are cream, widely spaced and thick, while the stem is cream, fibrous and very tough. **Spores** white, 6–7 x 3–4.5µm. Often in large troops under the shade of pinewoods in the south and east. Edibility uncertain.

Collybia maculata
(A. & S. ex F.) Kumm.
Cap 2–4in/5–10cm. This all-white
mushroom is very tough and fibrous,
especially the stem which may be
rather rooting in the soil. As it ages
the cap develops spots and streaks of
rust-brown. The taste is rather bitter.
Spores pale pinkish-buff, 5–6 x
4–5μm. Often in groups in the leaf-
litter of mixed woods, widespread.
Edible but tough.

Collybia butyracea
(Bull. ex Fr.) Quelet
Cap 2–3in/5–7.5cm. The
domed cap is smooth and
very greasy-waxy to the
touch, it also becomes
paler as it loses moisture
and is often two-toned.
The tough stem is club-
shaped and fibrous. The
crowded gills are often
slightly jagged on the
edges. **Spores** cream-
buff, 6–8 x 3–3.5μm.
Common under pines
and widely distributed.
Edible.

Collybia dryophila
(Bull. ex Fr.) Kumm.
Cap 1–3in/2.5–7.5cm. One of the
most common fungi everywhere,
the pale yellow-brown to reddish
cap is smooth and dry. The
crowded gills are whitish while the
smooth stem is reddish-brown
below with white or yellow hairs
at the base. **Spores** pale cream,
5–6 x 2–3μm. Often abundant in
mixed woods, especially under
oaks and pines, throughout
America. Edible.

Collybia alkalivirens
Singer
Cap ½–1½in/1–3cm. A very
dark blackish-brown to
reddish-brown all over, the
cap is radially furrowed, gills
are broad and rather distant.
The stem is fibrous-brittle
and hollow. All parts turn
deep green with a drop of
KOH (caustic soda). **Spores**
white, 5.4–8.5 x 2.2–5.4μm.
Frequent in mixed woods
throughout America.
Inedible.

Collybia confluens
(Pers. ex Fr.) Kumm.
Cap 1–2in/2.5–5cm. The
dull gray-brown to
pinkish caps and pale
brown stems, which are
densely hairy and grow in
large clusters, are the
principal characters to
look for. The gills are
narrow and very crowded.
Spores white, 7–10 x
2–4μm. On fallen leaves
or needles in mixed
woods, widely distributed.
Edible but poor.

Mycena galopus
(Pers.) Kummer
Cap ¼–½in/0.5–1.5cm. The small.
bell-shaped cap is grayish-black,
fading to whitish-gray. The gills are
pale gray while the stem is darker,
gray-brown below. When broken the
stem exudes a drop of white, milky
fluid. **Spores** white, 10–14 x 5–7μm.
A common species in mixed woods,
especially coniferous, widely
distributed. Inedible.

Mycena galericulata

(Scop. ex Fr.) S.F. Gray

Cap 1–2in/2.5–5cm. The broadly bell-shaped cap is grayish-brown to buffy-brown and often radially wrinkled. The gills are broad, white flushed grayish-pink and often with cross-veins between. The smooth stem is gray also, and often deeply rooting. **Spores** white, 8–11 x 5.5–7μm. This often very common species grows in tufts on rotting deciduous logs and stumps throughout North America. Edible but not recommended.

Mycena inclinata

(Fr.) Quel.

Cap 1–2in/2.5–5cm. Often confused with *M. galericulata* but the stem shades from reddish-brown below to yellowish above and has minute whitish flecks below (use a hand lens for this). The cap margin is often minutely toothed. The odor is distinct, fragrant, soapy to slightly rancid. **Spores** white, 8–10 x 5.5–7μm. Often very common on deciduous tree stumps in the north-east. Edibility doubtful.

Mycena haematopoda

(Pers. ex Fr.) Kummer

Cap ½–2in/1–5cm. The dark reddish-brown to wine-colored cap and stem are distinctive as is the deep reddish-brown juice which "bleeds" from any broken stem. **Spores** white, 9–10 x 6.5–7μm. A common species, growing in clumps on deciduous wood throughout North America. Edibility doubtful.

Mycena pura

(Pers. ex Fr.) Kummer
Cap 1–2in/2.5–5cm.
Cap, gills and stem range
from a delicate lilac or
pale violet to pinkish or
even blue-gray. What
remains constant is the
strong odor and taste of
radish. **Spores** white, 5–9
x 3–4µm. A common
species, usually growing
singly on leaf-litter,
throughout North
America. Inedible;
reported to cause
poisoning.

Mycena subcaerulea

(Peck) Sac.
Cap ¼–¾in/0.5–2cm. This
exquisite little mushroom has a
pale blue to greenish-blue cap
fading to brown at the center
with whitish gills. The stem is
colored like the cap. **Spores**
white, 6–8µm, globose. Quite
common, growing singly on
decaying leaf-litter of deciduous
trees east of the Great Plains.
M. amicta can be found in the
north-west, also blue, but
growing under conifers. Inedible.

Mycena leiana

(Berk.) Sacc.
Cap ½–2in/1.5–5cm. No other small
clustered mushroom on wood has
these brilliant orange, sticky caps, gills
and stems. The edges of the gills are a
darker reddish-orange. **Spores** white,
7–10 x 5–6µm. Common pushing
through cracks in the bark of fallen
deciduous trees and stumps, in
northern and eastern states. Inedible.

Mycena sanguinolenta
(A. & S. ex Fr.) Kummer
Cap ¼–½in/0.5–1.5cm.
Another species which
"bleeds" reddish juice when
broken, but this species is
usually single, not clustered,
and always very small, unlike
M. haematopoda, also shown
here. The colors are bright
reddish-orange to reddish-
brown. **Spores** white, 8–11
x 4–6µm. Common on
decaying leaf or pine needle
litter, throughout North
America. Inedible.

Mycena luteopallens
(Peck) Sacc.
Cap ¼–⅝in/1–1.5cm. This
tiny yellow-capped mushroom
has a slightly paler stem and
gills, and the stem base is
clothed in long hairs. If you
follow the stem down into the
soil it will always be found to
be attached to the old shells of
hickory nuts or walnuts.
Spores white, 7–9 x 4–5.5µm.
Often abundant but easily
overlooked, throughout
eastern states.

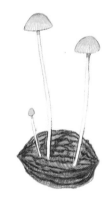

AMANITA FAMILY

There are over 120 species of *Amanita* in North America and the genus contains some of the most beautiful, and the most deadly, mushrooms in the world. Features to look for are the white spores, the universal veil which may be left as a sac-like volva at the stem base or as mere warty remnants on the cap and stem base. Also a partial veil may be present as a skirt-like ring or annulus at the top of the stem. Always make sure to collect the entire stem base from under the ground so any volva can be seen. Many *Amanitas* have distinctive odors. The gills are quite free from the stem as shown in the cross-sections below.

Amanita phalloides Death Cap
(Fr.) Sacc.

Cap 2–6in/5–15cm. This appears to be an introduction from Europe and is spreading rapidly in both western states and in the north-east. A deadly species it has already caused many poisonings. The cap varies from olive-green to yellowish or brownish. The older mushrooms smell of rotten honey.
Spores white, 8–10.5 x 7–8μm, non-amyloid. It grows under both oaks and pines and may fruit in very large numbers.

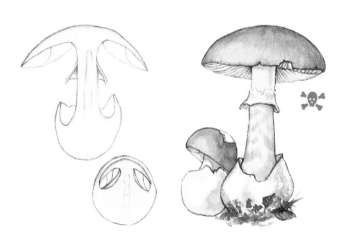

Amanita virosa Destroying Angel

Secretan.

Cap 2–8in/5–20cm.
Found in the Pacific
Northwest and extremely
common in north-eastern
states, this is another
deadly species which
poisons and kills many
people each year. The
pure white cap, stem, ring
and volva are distinctive
as is the pungent, sweet
odor. KOH (caustic soda)
on the cap turns it bright
yellow. **Spores** 9–11 x
7–9µm, amyloid. Other,
similar white species are
only separable
microscopically.

Amanita caesarea Caesar's Mushroom

(Scop. ex Fr.) Pers.

Cap 2–8in/5–20cm. The true *A. caesarea* is found under pines
in Arizona and New Mexico and is well known in the
Mediterranean region of Europe.The bright orange cap, thick
white volva and yellow gills are distinctive. **Spores** 8–12 x
6–8µm, non-amyloid. What is usually called the Caesar
Mushroom in north-eastern states is *A. hemibapha* (illustrated
p.84). This is a highly regarded edible, but as with all *Amanita*
species is not recommended in case of confusion with
poisonous species.

Amanita calyptrata Coccora Mushroom
Peck

Cap 2–10in/5–25cm. Found under pines, oaks and Madrones (*Arbutus* sp.), this species is widespread through northern California, Oregon, Washington, and Vancouver Island. The often very large caps vary from dark orange-brown, bronze, to paler yellowish-brown. The volva is very large and thick while the gills are white to pale creamy-yellow. **Spores** 8–11 x 5–6μm, non-amyloid. Highly regarded edible but not recommended in case of confusion with other species.

Amanita hemibapha Eastern Caesar's Mushroom
Bk. & Br. = *A. caesariea* of some authors

Cap 2–6in/5–15cm. The spectacular bright reddish-orange, umbonate cap, yellow-orange gills and stem with a floppy ring, emerging from a white volval sac make this an easily recognized species. **Spores** white, 8–12 x 6–8μm, non-amyloid. Found often in numbers, especially under oak and pine, in many eastern and southern states but with a curiously patchy distribution. Edible but like all *Amanita* not recommended. Compare with the true *A. caesariea* shown on p. 83.

Amanita brunnescens The Cleft-foot Amanita
Atkinson

Cap 2–6in/5–15cm. The brownish cap is often irregularly streaked with darker fibers and sometimes with decolored patches. The free gills are white and closely spaced. The stem is white, often discolored reddish-brown from below up, and the base is rather abruptly swollen with the edge split or cleft vertically. **Spores** white, 7–10µm, globose, amyloid. This common species is found in mixed deciduous woods, especially oak, southward through eastern states. Possibly poisonous. An all-white variety of this mushroom, *pallida*, can also be found, but it also has the cleft foot.

Amanita citrina
Schaeff. ex S.F. Gray

Cap 2–4in/5–10cm. This species comes in two color forms, a lovely pale lemon-yellow or with pale lavender tints in the large bulb and veil remnants. The rounded cap often has numerous patches of veil which once enclosed the young mushroom. The bulb at the stem base is large, rounded, with a distinct margin or gutter around the upper edge. The gills have a strong odor of freshly dug potatoes or radish and this forms an easy field character to recognize. **Spores** white, 7–10µm, globose, amyloid. Frequent in mixed woods in eastern North America. Inedible, easily confused with other deadly species.

Amanita porphyria
(A. & S. ex Fr.) Secr.

Cap 1–3in/2.5–7.5cm. The grayish-brown cap has faint purplish tints although these are more obvious on the stem and the ring. The stem ends in a large, rounded bulb which has a margin on the upper edge. The gills are white and free. The odor may be of potato or radish. **Spores** white, 7–9μm, globose. A fairly uncommon species found in mixed woods in the Pacific Northwest and through eastern states. Possibly poisonous.

Amanita vaginata
(Bull. ex Fr.) Vitt.

Cap 2–4in/5–10cm. This elegant species is one of the "ringless" *Amanita* species, sometimes called *Amanitopsis*. The cap is a delicate pale gray to steel-gray. At the base of the slightly grayish stem is a tall, white, volval sac. **Spores** white, 9–12μm, globose, non-amyloid. Frequent in mixed woods throughout North America. Edible but best avoided.

Amanita fulva Tawny Grisette
(Schaef. ex Fr.) Pers.

Cap 2–4in/5–10cm. The rich tawny to reddish-brown cap and long, elegant stem emerge from a white volval sac. There is no ring on the stem. The gills are white to cream and crowded. **Spores** white, 8–10μm, globose, non-amyloid. Common in mixed woods throughout North America. Edible but best avoided.

Amanita ceciliae

(B. & Br.) Bas = *A. inaurata*
Cap 2–5in/5–12.5cm. The dark
brown cap usually has many darker
fragments of veil left on the surface.
The gills are white to grayish and the
tall grayish stem has a gray, sac-like
veil which soon breaks into small
fragments, often left behind in the
soil. **Spores** white, 11.5–14μm,
globose, non-amyloid. A common
species in mixed woods throughout
North America. Edibility uncertain.

Amanita pachycolea

Stuntz
Cap 2–8in/5–20cm. This large,
ringless *Amanita* has a very dark,
almost black cap when young, then
soon becomes browner with age. The
margin is very deeply grooved and is
often much paler than the center of
the cap; there may also be a darker
zone just behind the pale band. The
stem is grayish-brown and has a thick,
white volval sac. The gills are soon
pale orange-buff. **Spores** white,
11–14.5 x 10–12.5μm, almost round,
non-amyloid. A frequent species
under pine and oak in the Pacific
Northwest. Edible but best avoided.

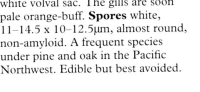

Amanita sinicoflava

Tulloss
Cap 2–3in/5–7.5cm. Although
officially described only very recently,
this has proved to be a common
species in many parts of the north-
east. The combination of yellow-
ochre to olive-ochre cap, pale buff
stem, and volval sac becoming
distinctly gray are very distinctive.
Spores white, 7–10 x 6–8.5μm non-
amyloid. Found in mixed woods,
especially under conifers. Edibility unknown.

Amanita muscaria Fly Agaric
Hooker (L. ex Fr.)

Cap 3–10in/7.5–25cm. This bright-red mushroom (and its yellow variety *formosa*) is probably the most well-known fungus in the world, appearing in numerous children's books and fairy tales. The white spots on the cap are remains of the universal veil. The stem and gills are white, while the stem has rings of white to yellow warts at the swollen base. A prominent, floppy ring is present. **Spores** white, 9.5–13 x 6.5–8.5μm, non-amyloid. The red form is common in the western states under pine and birch, the yellow form is very common in eastern states under pine. Poisonous; was traditionally used to poison houseflies, hence the common name.

Amanita pantherina
(DC. ex Fr.) Secr.

Cap 2–4in/5–10cm. The brownish cap has pure white fragments of veil and the gills and stem are also white. The abruptly bulbous stem has a narrow floppy ring above and one or more rings of veil tissue just above the veil. **Spores** white, 8–14 x 6.5–10μm, non-amyloid. Common in the west, but rarer in the east, usually under conifers. Very poisonous, causing delirium and coma-like sleep.

Amanita gemmata
(Fr.) Gill.

Cap 2–4in/5–10cm. The buffy-yellow cap has white patches of veil and the gills and stem are white. The base of the stem is bulbous with an abrupt margin or "gutter" on the upper edge. **Spores** white, 8.5–11 x 5.5–8.5μm, non-amyloid. Frequent in oak and pine woods, throughout North America. Possibly poisonous.

Amanita rubescens
(Pers. ex Fr.) S. F. Gray
The Blusher

Cap 2–6in/5–15cm. The common name refers to the pinkish-red blush that occurs wherever the fungus is bruised or damaged, and with age (splitting the stem usually shows reddish worm holes). The cap varies from yellowish-brown to reddish-brown. **Spores** white, 8–10 x 5–6μm, amyloid. Very common in mixed woods, widely distributed in North America. Edible but best avoided.

Amanita flavorubens
Atkinson

Cap 2–6in/5–15cm. Sometimes confused with the Blusher, *A. rubescens* because it also flushes pinkish-red where damaged or old, but the cap has strong yellow to yellow-brown colors. **Spores** white, 7.5–10 x 5.5–6.5μm, amyloid. Usually under oaks, quite common from Canada through eastern states, south to Florida. Edibility uncertain, best avoided.

Amanita spissa
(Fr.) Kumm.

Cap 2–6in/5–15cm. The grayish cap has fine, irregular grayish remnants of veil. The gills are white while the stem is white with grayish overtones. The base of the stem is swollen and slightly rooting. The odor is often of radish.
Spores white, 9–10 x 7–8μm, amyloid. Uncommon in mixed woods in eastern states. Edibility uncertain, best avoided.

Amanita francheti

(Boud.) Fayod = *A. aspera*

Cap 2–6in/5–15cm. This essentially brown-capped mushroom is covered in bright yellow fragments of veil and these fragments are also found on the base of the stem and on the edge of the ring at the stem apex. **Spores** white, 8–10 x 6–7µm, amyloid. Rather uncommon, under conifers in the Pacific Northwest and California. Edibility uncertain, best avoided.

Amanita abrupta

Peck

Cap 2–4in/5–10cm. An all-white species distinguished by the slender stature with a very abrupt, bulbous base to the stem. The cap surface has small, firm warts and the margin has small, dangling fragments of veil. The ring is large, floppy and has small "teeth" at the edge. **Spores** white, 6.5–9.5 x 5.5–8.5µm, amyloid. Particularly common in the south-east but found as far north as Vermont, in mixed woods. Edibility unknown.

Amanita onusta

(Howe) Sacc.

Cap 1–4in/2.5–10cm. The gray cap is covered in small, conical to irregular warts, powdery at the margin. The gills are whitish, crowded. The stem is often deeply rooted in the soil with a pointed bulb, and is sprinkled with gray, warty veil remnants. The odor is odd, like bleach, or old ham. **Spores** white, 8–11.5 x 5–8.5µm, amyloid. Quite common in sandy, loose soils in mixed woods throughout eastern and south-eastern states. Inedible.

Amanita longipes
Bas
Cap 1–3in/2.5–7.5cm.
This all-white species is
distinctive by its very long,
slender, and deeply rooting
stem. The cap and stem
have a loose, powdery
white covering of veil
fragments. There is no true
ring present, just a ring-
zone. The odor is nil to
faint of bleach. **Spores**
white, 8.5–17.5 x 4–7.5μm,
amyloid. Rather
uncommon, in mixed
woods in the north- and
south-east. Inedible.

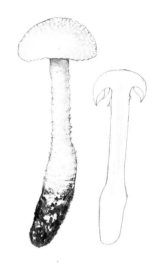

Amanita daucipes
(Mont.) Lloyd
Cap 3–10in/7.5–25cm.
This magnificent fungus is
easily recognized by its pale
salmon-pink colors and
very large, turnip-like basal
bulb. The edge of the bulb
often has small teeth-like
fragments of veil. The odor
is strong of old ham. The
ring on the stem is very
large and soft, like cream
cheese in texture and soon
collapses. **Spores** white,
8.5–11.5 x 5–7.5μm. Quite
common in late summer in
deciduous woods
throughout eastern states.
Inedible.

Amanita cokeri
(Gilb. & Kuhner) Gilbert
Cap 3–6in/7.5–15cm.
Another all-white species
with soft warts on the
cap, a large bulb which
has rings of recurved
scales at the top, and a
floppy ring. There is
sometimes a secondary
ring below the first.
Spores white, 11–14 x
7–9.5μm, amyloid. Quite
common in mixed woods,
in the east and south-
eastern states. Inedible.

Amanita volvata
(Peck) Lloyd
Cap 2–4in/5–10cm. The
white cap soon develops
pinkish tints and usually
has a mosaic of flattened
veil fragments. The gills
are whitish, crowded. The
stem has no ring and has
a thick, white volval sac at
the base. **Spores** white,
8.5–10 x 5.5–7μm,
amyloid. Quite common
in mixed woods
throughout the north-
east. Inedible. This
appears to be a complex
of 2 or 3 very similar
species distinguished
microscopically.

Amanita crenulata
Peck

Cap 1–4in/2.5–10cm. The cap
is a pale brownish-beige to
yellowish-tan and has small,
often conical warts in irregular
patches. The gills are white.
The stem is bulbous with a
narrow rim of small granular
veil fragments. A ring is present
but is soon torn and sometimes
collapses. **Spores** white, 7.5–10
x 6–8.5μm, non-amyloid.
Poisonous, causes delirium,
sweating, violent fits and finally
a coma-like sleep. Apparently
quite common in some areas of
the north-east.

Amanita flavoconia
Atkinson

Cap 1–3in/2.5–7.5cm. One of the most common species and
one of the loveliest; features are the bright yellow-orange cap
and yellow fragments of veil, white to yellowish gills, the
white, bulbous stem which has crumbly patches of bright
yellow veil at the base (often left in the soil when picked).
Spores white, 7–9.5 x 4.5–5μm, amyloid. Often abundant in
mixed woods in eastern North America. Inedible. *A. frostiana*,
almost identical, differs in the edge of the cap being radially
lined, the often marginate bulb
and non-amyloid spores.

Amanita parcivolvata
(Peck) Gil.
Cap 1–4in/2.5–10cm. This elegant
species has a bright orange-red cap
with yellowish veil patches, yellowish
gills and a bulbous stem which has a
powdery yellow coating. There is no
ring on the stem. **Spores** white,
10–14 x 6.5–8μm, non-amyloid.
Rather uncommon, in mixed woods,
especially oaks, from New Jersey
down to Florida, west to Ohio.
Inedible.

Amanita wellsii
Murrill
Cap 2–5in/5–12.5cm. A very unusual
color in *Amanita*, a pale salmon-pink
cap with dull yellowish patches. The
gills are white and the bulbous stem is
pale yellow with a yellow, very fragile
ring which tears and mostly clings to
the edge of the **cap. Spores** white,
11–14 x 6.5–8.5μm, non-amyloid. In
mixed woods from New York up to
Canada, also in the mountains of
North Carolina. Inedible.

Limacella glioderma
(Fr.) Maire
Cap 1–3in/2.5–7.5cm. The very slimy
cap is dark reddish-brown to pinkish
at the margin, the center is often
granular under the slime. The gills are
white and free while the stem is light
brown with a very slimy veil over the
surface. The odor is strong, mealy.
Spores white, 3–4μm, globose, non-
amyloid. A rather rare species, found
under hemlock and birch, widespread.
Edibility unknown, best avoided. A
close relative of the genus *Amanita*
differing principally in the glutinous
veil, although some mycologists think
it is closer to the following genus, *Lepiota*.

LEPIOTA FAMILY

These mushrooms have gills free from the stem, white to pale pinkish-cream, or green spores and usually a more or less obvious ring on the stem. They are closely related to the black-brown spored *Agaricus* mushrooms.

Lepiota cristata
(Fr.) Kumm.

Cap ½–2in/1.5–5cm. The small, white cap is marked with concentric rings of dark, reddish-brown scales. The white gills are free of the stem and the slender white stem has a tiny white ring towards the top. The odor is strong, unpleasant, of rubber or of the common Earthball *Scleroderma*. **Spores** white, 5–7 x 3–4μm, wedge-shaped, dextrinoid. Frequent in grassy areas in woods and pathsides throughout America. Possibly poisonous.

Lepiota clypeolaria
(Bull. ex Fr.) Kumm.

Cap 1–3in/2.5–7.5cm. The cap has small, brownish scales while the yellowish-brown stem is very shaggy-wooly up to a slight ring-zone. The free gills are white. **Spores** white, 12–20 x 4–5μm, spindle-shaped, dextrinoid. In mixed woods throughout North America. Poisonous.

Lepiota rubrotincta
Peck

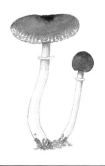

Cap 1–3in/2.5–7.5cm. The cap is a lovely coral-red and minutely scaly towards the edge, it usually has a darker center. The gills and stem are white, and there is a small ring at the top of the stem. **Spores** white, 6–10 x 4–6μm, elliptical, dextrinoid. In deciduous woods throughout North America. Edibility unknown.

Lepiota josserandii
Bon & Boif.

Cap 1–2in/1.5–2.5cm. The cap has bands of concentric brown scales and the cinnamon brown stem is scaly up to the small ring-zone. The gills are free, whitish to straw-yellow. There is often a rather unpleasant, musty odor. **Spores** white, 6–8 x 2.5–4.5μm, elliptical, dextrinoid. In deciduous woods and gardens, western North America. Deadly poisonous.

Lepiota americana
Peck

Cap 2–6in/5–15cm. The reddish-brown cap is broken up into very fine scales showing the white underlayer near the margin. The stem is swollen, spindle-shaped and rooting, white staining yellow then reddish-brown when bruised or cut. There is a fragile ring on the stem. **Spores** white, 8–14 x 5–10μm, elliptical, dextrinoid. On or around old stumps or on sawdust and woodchips, eastern North America, common. Edible but best avoided.

Leucocoprinus birnbaumi
(Corda) Singer = *Lepiota lutea*

Cap ½–1in/1.5–2.5cm. This beautiful, fragile, all-yellow mushroom is found only in greenhouses or in pots of houseplants in northern states, but outdoors in leaf-litter or compost in the south. **Spores** white, 8–13 x 5–8μm, elliptical, dextrinoid. This species probably originates from the tropics. Inedible.

Macrolepiota procera

(Scop.) Singer = *Lepiota procera*
Cap 5–10in/12.5–25cm.
This very tall, stately
species is one of our largest
mushrooms. The brown,
coarsely scaly cap, stem
with brown banding, and
thick, double-edged ring
are features to look for.
The gills are white and
free. **Spores** white to
cream, 15–20 x 10–13µm,
elliptical. Found in open
meadows, woodland edges
and roadsides, mostly in
north-eastern states. Edible
and delicious.

Macrolepiota prominens

(Fr.) Moser = *Lepiota prominens*
Cap 4–6in/10–15cm. Even taller
and more graceful than the better
known *M. procera*, this species is
striking for its almost white cap and
stem with only faint brown
markings. The cap scales are also
much finer. The ring is double
edged and thick. The gills are cream
to slightly pinkish. **Spores** pinkish-
cream, 9–10 x 6–7µm, elliptical. In
woodland clearings, fields and
meadows, often quite common in
the north-east. Edible and delicious.

Macrolepiota gracilenta
(Fr.) Wasser = *Lepiota gracilenta*

Cap 3–4in/7.5–10cm.
Slender, graceful, this species
has a finely scaly, pale brown
cap and pale brown stem with
faint banding. The ring is thin,
often funnel-shaped when
young. The gills are white.
Spores cream, 10–13 x
7–8μm, elliptical. This is the
most common large *Lepiota* in
eastern North America and is
often mistaken for *M. procera*.
It grows in wooded areas
rather than in open fields.
Edible and good.

Macrolepiota rhacodes
(Vitt.) Singer

Cap 3–6in/7.5–15cm. The shaggy brown cap, bulbous stem
which bruises deep reddish-brown, and the thick, double ring
are the distinctive features. **Spores** white, 6–10 x 5–7μm,
elliptical. In woods in deep leaf-litter, or in gardens on
compost or heaps of leaves, common throughout North
America. Edible and delicious. The similar, but larger
M. venenata Bon grows in Southern California; it has a large,
marginate bulb, coarse scales and is poisonous.

Leucoagaricus naucina

(Fr.) Singer

Cap 2–4in/5–10cm. At first all-white, this smooth-capped species eventually ages to a dull grayish-white, The gills are white then grayish-pink. The stem has a small, double-edged ring at the top. **Spores** at first white, but later deposits are pink, 7–9 x 5–6μm, oval. In open fields, roadsides and lawns, common throughout North America. Has caused some stomach upsets, best avoided.

Chlorophyllum molybdites

Massee

Cap 4–12in/10–30cm. Often huge in size this beautiful species has a snow-white cap with darker buffy scales at the center. The gills are at first white then age to a dull gray-green. The stem is bulbous, white, and has a thick, double-edged ring. **Spores** green, 8–13 x 6.5–8μm, elliptical. Often in huge circles in open fields and lawns, common in the south and north to New Jersey and New York. Poisonous, causes severe stomach upsets, always check the gill and spore color on large *Lepiotas*.

BLACK, DEEP BROWN TO PURPLE-BROWN SPORES

GOMPHIDIUS FAMILY

Although these mushrooms have gills they are related to the boletes with their spongy pores because of their very similar spores and other microscopic characters. Most species have a veil which may be dry or glutinous.

Gomphidius subroseus
Kauffman

Cap 2–4in/5–10cm. The beautiful rose-red to deep red cap, whitish to gray decurrent gills and white stem, and often slimy surface are easy features to recognize, as is the habitat under conifers, especially Douglas-fir. **Spores** blackish, 15–20 x 5–7μm. It is widely distributed in northern and western states and in the mountains of states further south. Edible.

Chroogomphus glutinosus
(Schaeff. ex Fr.) Fr.

Cap 2–4in/5–10cm. The pallid, gray-brown to purplish-brown cap is slimy, the decurrent gills are grayish-white and the stem is white with a yellow base and smeared with a slimy veil. **Spores** blackish, 15–21 x 4–7.5μm. Frequent throughout North America under conifers, especially spruce. Edible but not recommended.

Chroogomphus vinicolor
(Pk.) O.K. Miller

Cap 1–3in/2.5–7.5cm. *Chroogomphus* differs from *Gomphidius* in its pinkish-yellow, rather than white, flesh. This rather small species is distinguished by its cap which is ochre flushed with wine-red and has a small, central, pointed knob. The gills are decurrent, thick and grayish-ochre. **Spores** 17–23 x 4.5–7.5µm. It is widely distributed in North America under pines and is edible.

AGARICUS FAMILY

Agaricus provides us with the well-known mushroom we purchase in stores, also the popular field mushroom. All species have deep brown spores, free gills, and a veil forming a ring on the stem.

Agaricus campestris Field mushrooms
L. ex Fr. & *A. andrewi* Freeman

Cap 2–4in/5–10cm. These two species are so similar they may be treated as if one species. They have a white to grayish cap which may become fibrous-scaly with age. The gills start bright pink then mature to dark brown, and the short, fat stem is white with a fine, fragile ring. The flesh is white, turning slightly reddish when cut. **Spores** brown, 6–9 x 4–6µm. Often abundant and in fairy rings in lawns and meadows throughout North America. *A. andrewi* seems to be the commoner species in the north-east and differs by having large, turnip-shaped cells (cystidia) on the gill edges. Both are edible and delicious.

Agaricus cupreobrunneus

(Moll.) Pilat
Cap 1–3in/2.5–7.5cm. The copper-
brown cap is slightly fibrous-scaly.
The gills start bright pink then mature
deep brown. The short stem is white
with a fragile, thin ring. **Spores**
brown, 7–9 x 4–6.5μm. In lawns,
meadows, common in California and
other western states. Edible and good.

Agaricus bitorquis

(Quel.) Sacc. = *A. rodmani*
Cap 2–6in/5–15cm. The flattened
white cap has extremely thick flesh
compared with the very narrow
gills. The latter start pinkish and
turn deep brown. The stem is
stout with a curious double ring
on the lower stem, almost like a volva.
Spores brown, 5–6 x 4–5μm. This
species often pushes up through hard
asphalt or gravel along roads, and
prefers packed ground in urban areas.
Edible and good.

Agaricus subrutilescens

(Kauff.) Hot. & Stuntz
Cap 2–6in/5–15cm. The
fibrous-scaly cap is
reddish- to purplish-
brown. The gills start
white then become
pinkish-brown to deep
brown. The club-shaped
stem has a well-
developed ring and is
wooly-fibrous below. The
odor is fruity. **Spores**
brown, 5–6 x 3–4μm.
Solitary to grouped in
conifer woods along the
west coast. Has caused
stomach upsets, best
avoided.

Agaricus augustus
Fries

Cap 4–10in/10–25cm. This stately mushroom has a scaly-fibrous, golden-brown cap. The gills are whitish-pink then dark brown. The tall stem has a large floppy ring and is very wooly-scaly below. When bruised it turns dull yellow. The odor is pleasantly anise. **Spores** brown, 8–11 x 5–6μm. Quite common especially under redwoods near paths and roads in the Pacific Northwest and other western states. Edible and choice.

Agaricus arvensis Horse Mushroom
Schaeff. ex Secr.

Cap 4–6in/10–15cm. The cap is smooth then slightly scaly with age, the gills start white then turn deep brown. The stout stem is white and smooth and has a large, floppy ring which has thick, toothed or cog wheel-like tissue on the underside. The odor is pleasant of almond-anise. **Spores** brown, 7–9 x 4.5–6μm. In fields and woodland clearings, often under spruce, throughout North America. Edible and delicious.

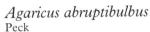

Agaricus abruptibulbus
Peck

Cap 3–6in/7.5–15cm. An all-white species and rather elegant, the gills start pale pink then mature brown. The stem has a prominent, very abrupt bulb. When bruised all parts turn a dull yellow and the flesh has a pleasant odor of anise. **Spores** brown, 6–8 x 4–5μm. Common in mixed woods in the north and north-east. Edible but must not be confused with deadly white *Amanita* species.

Agaricus silvaticus
Schff. & Secr.
Red Staining Mushroom
Cap 2–4in/5–10cm. The scaly cap
is bright reddish- to yellow-brown while
the gills are pink then dark brown.
The stem is slender, bulbous with a
white ring, smooth on the underside.
The flesh when cut turns more or less
bright red. **Spores** brown, 4.5–6 x 3–3.5μm.
In conifer woods in the Pacific Northwest.
Edible. There are other red-stainers such as
A. fuscofibrillosus with fibrillose (not scaly)
reddish-brown cap and *A. benesi*, an all-white species.

Agaricus placomyces
Peck
Cap 2–4in/5–10cm. This species is
rather slender with a finely fibrous-
scaly blackish-brown cap. The gills
are whitish-pink then brown. The bulbous
stem is white with a floppy ring that often
has yellowish droplets on the underside.
The odor is unpleasant, like iodine. **Spores**
brown, 5–7 x 3–4μm. Common in
mixed woods in the north-east. Causes
stomach upsets. The very similar *A. pocillator*
has a small but prominent bulb with a rim and
has a double ring.

Agaricus praeclaresquamosus
Freeman
= *A. meleagris* of American authors
Cap 2–6in/2–15cm. The finely scaly
cap is dark grayish-brown. The gills
start whitish-pink then turn dark
brown. The stem is smooth below the
prominent ring which is thick, felt-like
and often split at the margin. The
odor of the flesh when crushed is
unpleasantly of phenol or iodine.
Spores brown, 4–6.5 x 3–4.5μm.
Frequent along roads and paths under
trees in the western states. Poisonous
to many, causing nausea and vomiting.

Agaricus xanthodermus Yellow Stainer
Gen.

Cap 3–5in/7.5–12.5cm. The white cap is often fibrous-scaly and turns dull grayish. The gills are whitish-pink then brown. The bulbous stem is white with a well-developed ring with cottony patches on the underside. The odor of the flesh when rubbed is strong, unpleasantly of iodine or old-fashioned school ink. When cut, the flesh, especially at the base of the stem, turns bright, chrome-yellow. **Spores** brown, 5–7 x 3–4μm. Frequent in urban areas by paths, hedges and in mixed woods, in western states. Poisonous to many causing nausea, headaches.

Agaricus micromegathus
Peck

Cap 1–2in/2.5–5cm. This small species has a pale yellowish-brown cap which is finely scaly and very faintly flushed with purple on occasion. The gills are broad and pink before turning pale brown. The short, sturdy stem has a faint ring-zone. All parts bruise yellow on handling, especially the stem. **Spores** brown, 4.5–5.5 x 3.5–4μm. Found in short turf in lawns and fields, in eastern states. Edibility uncertain.

STROPHARIA FAMILY

The gills are sinuate to adnate, purplish-brown, spores are deep purple-brown. The stem usually has a ring and the cap may be dry to glutinous.

Stropharia rugosoannulata
Farlow ex Murr.

Cap 4–6in/10–15cm. A very variable mushroom depending on its age and exposure to sun, it will change from deep wine-brown to a light tan. The gills start pale grayish-lilac and turn purple-black. The tough, white stem is smooth with white "roots" at the base (rhizomorphs) and with a thick, toothed or segmented ring above. **Spores** purple-brown, 10–13 x 7.5–9μm. Common on woodchip mulches in gardens, often in hundreds, starting in the spring and early summer, widespread in North America. Edible and good and often cultivated.

Stropharia thrausta
(Schulz.) Sacc.

Cap 1–2in/2.5–5cm. An uncommon but beautiful species with bright reddish-orange colors on cap and stem, with small whitish scales present on the cap. **Spores** purplish-black, 12–14 x 6–7μm. Found on small twigs and fallen branches of deciduous wood, widely distributed. Inedible.

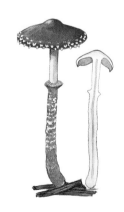

Stropharia semiglobata
(Batsch) Quel.

Cap 1–2in/2.5–5cm. The
rounded, pale yellow cap is
sticky when wet, while the
tall, slender stem is only
sticky below the faint ring-
zone. The gills are grayish-
lilac. **Spores** pale violet-
brown, 15–19 x 8–10µm.
Common on horse, sheep and
cattle dung everywhere.
Inedible.

Stropharia aeruginosa
(Curt.) Quel

Cap 1–3in/2.5–7.5cm. Truly
green mushrooms are rare and
this is one of the very few to be
found. The sticky cap has
prominent white flecks of veil
and is yellower at the center.
The stem is also flecked-scaly
with white on a green ground
up to a ring-zone. The gills are
deep violet-gray. **Spores** violet-
brown, 7–9 x 4–5µm. On rich
soil or in grass at edges of
woods, mostly in western states.
Inedible.

Psilocybe cubensis
(Earl) Singer

Cap ½–3in/1.5–7.5cm. The
conical or bell-shaped cap is sticky
and white with a brownish-yellow
center. The gills are crowded, deep
violet-black with whitish edges.
The slender stem is fibrous-
grooved at the top, white, and
bruises blue. **Spores** purple-
brown, 11.5–17 x 8–11.5µm. On
cow and horse dung in pastures in
the south and south-eastern states.
Inedible, hallucinogenic.

Psilocybe semilanceata Liberty Caps
(Fr ex Secr.) Kumm.

Cap ¼–1in/0.7–2.5cm. The narrowly
conical cap is smooth and sticky when
wet, pale yellow-buff to brown. The
narrow gills are grayish-brown and
crowded. The stem is very slender
and sinuous, whitish bruising blue
when handled. **Spores** purple-brown,
11–14 x 7–8µm. In grass in fields and
pastures, often abundant, in the
Pacific Northwest. Inedible,
hallucinogenic. The common name
refers to the resemblance to French
revolutionary hats.

Hypholoma fasciculare Sulfur Tuft
(Huds. ex Fr.) Kumm. =
Naematoloma fasciculare

Cap 1–3in/2.5–7.5cm. The tufted
growth, bright sulfur-yellow caps and
stems, and the dark purple-brown
spores are all distinctive features. The
young gills are greenish-yellow before
maturing to purple-brown. **Spores**
6.5–8 x 3.5–4µm. On logs, stumps
and buried wood, throughout North
America, common. Poisonous.

Hypholoma sublateritium Brick Caps
(Fr.) Quel. = *Naematoloma
sublateritium*

Cap 2–4in/5–10cm. These are
also in tufts but they are larger,
stouter and a deep brick-red.
The gills are whitish then
purple-gray. **Spores** purple-
brown, 6–7 x 4–4.5µm. On dead
stumps and logs, common in
eastern North America, late in
the year. Edible but must not
be confused with *H. fasciculare*.

COPRINUS FAMILY

Usually thin, delicate mushrooms (there are exceptions) they all have deep black or blackish-brown spores. Many species have a veil and/or a ring. *Coprinus* species are often called ink-caps because they dissolve away into an inky liquid as they disperse their spores.

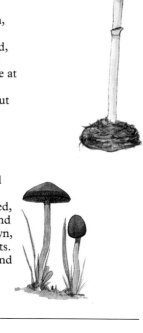

Panaeolus sphinctrinus
(Fr.) Quel.
Cap 1–2in/2.5–5cm. The distinctive feature is the tiny white "teeth" which hang down at the edge of the gray cap. The gills are narrow, grayish-black, mottled with tiny black spots. The stem is slender, gray. **Spores** blackish, 13–16 x 8–11µm. On horse or cow dung, widely distributed. Inedible.

Panaeolus semiovatus
(Sow. ex Fr.) Lund. & Nannf.
Cap 1–3in/2.5–7.5cm. The whitish, egg-shaped cap often cracks in dry weather. The gills are widely spaced, blackish-gray and mottled. The tall stem has a distinct ring or ring-zone at the top. **Spores** blackish, 15–20 x 8–11µm. On horse dung, throughout North America. Inedible.

Panaeolus foenisecii
(Pers. ex Fr.) R. Maire.
Cap ½–1in/1–2.5cm. The rounded cap starts date-brown but dries out pale buff. The gills are widely spaced, dark brown. The stem is slender, and pale buff. **Spores** dark purple-brown, 12–15 x 6.5–9µm, with minute warts. Very common, scattered in lawns and fields throughout North America whenever the weather is damp. Inedible.

Psathyrella candolleana
(Fr.) Maire
Cap 1–4in/2.5–10cm. This
very delicate mushroom breaks
with the least handling, the
brittle caps and stems are
buffy-brown then almost ivory-
white when dry. The gills are
narrow, grayish-lavender.
Spores purplish-brown, 7–10
x 4–5μm. Singly to small tufts
often on buried wood in grass
or by stumps. Widely
distributed in North America.
Edible.

Psathyrella velutina
(Pers. ex Fr.) Singer
Cap 1–3in/2.5–7.5cm. The
reddish-brown cap and stem
are both hairy-fibrous and
there is a hairy ring-zone at the
stem apex. The broad gills are
deep yellow to blackish-brown
with white edges, and often
weeping droplets. **Spores**
blackish-brown, 9–12 x
6–7μm. In woods and gardens,
usually in disturbed soil along
tracks and paths. Edible but
best avoided.

Psathyrella delineata
(Peck) Smith
Cap 2–4in/5–10cm. The curiously
wrinkled, corrugated cap is unusually
large and fleshy for this genus. It
starts a deep, watery brown but dries
paler. The gills are attached, close,
and purplish-brown. The stem has a
faint ring-zone at the top. **Spores**
purplish-brown, 7–9 x 4–6μm. Quite
common on dead, woody debris and
leaf-litter in the east. Edibility
unknown.

Coprinus quadrifidus
Peck

Cap 1–3in/2.5–7.5cm. The egg-shaped cap is up to 3in/7.5cm tall and has a surface covered with broken patches of veil. The narrow gills are white then inky-black. The white stem is scaly with a slight ring-zone. **Spores** black, 7.5–10 x 4–5µm. The cap dissolves away (deliquesces) as it matures. Uncommon, often in large clusters, on woody debris in eastern states. Edible but not recommended.

Coprinus comatus Shaggy Mane
(Mull. ex Fr.) S.F. Gray

Cap 1–3in/2.5–7.5cm across, 3–6in/7.5–15cm high. The cap is unmistakable, tall, cylindrical and shaggy white scales, all on a tall, straight stem with a ring low down. The entire cap dissolves away in a few hours. **Spores** black, 11–15 x 6–8.5µm. Frequent in disturbed soil in gardens and along road edges, throughout North America. Edible and good.

Coprinus atramentarius Common Ink Cap
(Bull. ex Fr.) Fr.

Cap 1–3in/2.5–7.5cm. The egg-shaped cap is smooth gray-brown with some slight scales at the center. The white, crowded gills mature black. The white stem has a ridge or ring-zone at the very base. **Spores** black, 7–11 x 4–6µm. As with all *Coprinus* it dissolves away when mature. Very common in clusters near stumps and buried wood in grass, throughout North America. Edible but reacts with alcohol to produce nausea, flushing and tingling.

Coprinus micaceus
(Bull. ex Fr.) Fr.

Cap 1–2in/2.5–5cm. The whole cap sparkles with minute, mica-like specks of veil when fresh, although they wash off with age. **Spores** black, 7–10 x 4–5µm. Often in very large clusters on or around stumps, throughout North America. Edible but worthless.

Coprinus plicatilis
(Curtis ex Fr.) Fr.

Cap ¼–1in/0.6–2.5cm. The deeply furrowed cap soon flattens and has a darker, brown center. The flesh is so thin you can almost see through the cap. The stem is extremely fragile and the mushroom only lasts a few hours. **Spores** black, 10–13 x 6.5–10µm. In grass throughout North America. Edible.

RUST-BROWN, YELLOW-BROWN, CIGAR-BROWN SPORES

PAXILLUS FAMILY

This group, despite their gills, are thought to be closely related to the boletes. They share many microscopic and chemical features with that group and have similar spore colors.

Paxillus involutus
(Bat. ex Fr.) Fr.

Cap 2–6in/5–15cm. An important species because it can be poisonous and even fatal in some cases. The distinguishing features are the yellow-brown cap with a wooly, at first inrolled, cap margin, and the soft yellow-brown gills which descend the stem and bruise deep reddish-brown. **Spores** brown, 7–9 x 4–6µm. Widespread throughout America but not especially common.

Paxillus atrotomentosus
(Bat. ex Fr.) Fr.

Cap 2–8in/5–20cm. This often very large mushroom is found on stumps of old pine trees and is easily recognized by the brownish cap with decurrent yellow gills, and thick, deep brown, velvety-hairy stem. **Spores** yellowish, 5–6 x 3–4µm. It is common over much of North America. It is supposedly edible but best avoided.

Phylloporus rhodoxanthus
(Schw.) Bres.

Cap 1–4in/2.5–10cm. This species has a velvety, reddish-yellow to reddish-brown cap, deep yellow gills which are often cross-veined and wrinkled and a reddish stem. **Spores** brown and 9–12 x 3–5µm. It is very common over much of North America in mixed woods. It is edible but poor in quality.

CORTINARIUS FAMILY

This is the largest group of mushrooms in the world and they grow in a bewildering number of forms, colors and sizes. The spores vary from rust brown to dull, cigar-brown and may be smooth, warty or even angular-lumpy. A cobwebby veil is often present.

Cortinarius trivialis
Lange

Cap 2–4in/5–10cm. The sticky cap is yellow-brown while the gills are pale lilac before turning rust-brown. The stem is sticky with rings of sticky, whitish to yellow veil. **Spores** rust-brown, 10–15 x 7–8µm, warty. Frequent in boggy places in deciduous woods, in the north and north-eastern states. Inedible.

Cortinarius muscigenus
Peck = *C. collinitus* of some authors, = *C. cylindripes*

Cap 2–4in/5–10cm. The sticky cap is rich orange-brown to ochre, smooth. The gills are white at first while the stem is flushed with pale blue-violet, and banded with thicker rings. **Spores** rust-brown, 12–15 x 7–8µm, warty. Usually under spruce, in wet woods in the north and north-east. Inedible.

Cortinarius pseudosalor
Lange

Cap 2–4in/5–10cm. The very slimy cap is pale ochre-brown and wrinkled while the gills are pale clay-buff. The slimy stem is cylindric, pale bluish-lilac. **Spores** rust-brown, 12–15 x 6–8µm, warty. In mixed woods in the north and east. Inedible.

Cortinarius delibutus
Fries

Cap 2–4in/5–10cm. Often very sticky the cap is bright golden-yellow and smooth. The gills and stem apex are pale lavender when young, the lower stem has a sticky yellow covering. **Spores** rust-brown, 7–8 x 5–6µm, warty. In damp, mossy woods, throughout the north and eastern states. Inedible.

Cortinarius iodes
Berk. & Curt

Cap 1–3in/2.5–7.5cm. Probably the most common of the all-violet species the cap is usually blotched and spotted with areas of white. Both cap and stem are sticky when wet but soon dry. **Spores** rust-brown, 8–12 x 5–6.5µm, warty. The very similar *C. iodeoides* differs in a bitter-tasting cap surface and smaller spores. Inedible.

Cortinarius volvatus
A.H. Smith

Cap 2–4in/5–10cm. An unusual species because the base of the stem has a thick white veil almost in the form of a volval sac. All parts of the mushroom are bluish when young but the cap turns brownish with age. The taste is bitter. **Spores** rust-brown, 8–10.5 x 5–6µm, warty. In mixed woods on calcareous soils, uncommon, in the north and north-east. Inedible.

Cortinarius calyptrodermus
A.H. Smith

Cap 2–4in/5–10cm. This beautiful species is striking for the very thick white patches of veil against the blue-lilac cap. The gills and bulbous stem are also bluish. **Spores** rust-brown, 10–13 x 6.5–8µm, warty. In mixed woods, apparently widely distributed in the north and east, but uncommon. Inedible.

Cortinarius elegantiodes
Peck

Cap 1–3in/2.5–7.5cm. The slightly sticky cap is brilliant orange as is the lower half of the bulbous stem. The gills are bright yellow. **Spores** rust-brown, 14–19 x 8–10µm, warty. In beech woods on calcareous soils in the northeast. Inedible. This has the largest spores of any orange, bulbous *Cortinarius*.

Cortinarius purpurascens
Fries

Cap 2–4in/5–10cm. The dark brown cap is streaked with darker fibers and is sometimes violet at the margin. The gills and stem are violet and both bruise deep violet-purple. The stem has a distinct marginate bulb. **Spores** rust-brown, 8–10 x 4–5.5µm warty. In coniferous woods, in the north-east, occasional. Inedible.

Cortinarius torvus
(Fr.) Fries
Cap 2–4in/5–10cm. A rather dull, brown species distinguished by the whitish veil on the lower stem which looks like a stocking half pulled up. The gills and stem apex are violet when young. **Spores** rust-brown, 8–10 x 5–6μm, warty. Common in beech and oak woods in the north-east. Inedible.

Cortinarius armillatus
(Fr.) Fr.
Cap 2–5in/5–12cm. One of the most common species, it often grows in hundreds. The reddish-brown, dry caps are smooth to fibrous. The gills are broad, distant and pale cinnamon-brown. The stem is often very bulbous and has 2–4 narrow bands of bright reddish veil. **Spores** rust-brown, 9–12 x 5.5–7.5μm, warty. In beech and conifer woods throughout the north-east. In the west C. haematochelis has a darker cap, reddish-brown belts, and smaller, rounder spores. Inedible.

Cortinarius evernius
(Fr.) Fr.
Cap 2–4in/5–10cm. The dark, reddish-brown, fibrous cap contrasts with the beautiful violet stem which is banded with white veil and is often long and spindle-shaped. **Spores** rust-brown, 9–11 x 5–6μm, warty. Common in wet, boggy woods under conifers, throughout the northern states. Inedible.

Cortinarius alboviolaceus
(Pers. ex Fr.) Fr.
Cap 2–5in/5–12cm. The specific name means whitish-violet which is exactly what this fungus is. The whole mushroom is silvery white with a flush of violet or lavender. The bulbous stem is ringed or booted with whitish veil at the base. **Spores** rust-brown, 8–10 x 5–6µm, warty. Common in mixed woods, throughout North America. Inedible.

Cortinarius argentatus
(Pers. ex Fr.) Fr.
Cap 2–4in/5–10cm. The whole mushroom is silvery white to slightly lilac, becoming yellowish at the center. The stem is often very bulbous and is almost smooth and without veil. The gills are pale brown when mature. **Spores** rust-brown, 8–10 x 5–6µm, warty. Quite common in early summer in beech woods in the north-east. Inedible.

Cortinarius bolaris
(Pers. ex Fr.) Fr.
Cap 2–3in/5–7.5cm. The yellowish cap becomes spotted with copper-red scales as does the stem also. The gills are pale buff becoming rust-brown. When cut or bruised all parts turn dull yellowish then eventually reddish. **Spores** rust-brown, 6–7 x 5–6µm, warty. Common in beech woods throughout north-east. Possibly poisonous.

Cortinarius violaceus
(Fr.) S.F. Gray

Cap 2–4in/5–10cm. Unmistakable by its deep, blackish-violet fruit-body with dry, minutely scaly cap and club-shaped stem. The violet gills are rust-brown when mature. **Spores** rust-brown, 13–17 x 8–10μm, warty. Rather uncommon, in beech woods in the north-east. The subspecies *hercynicus* (Pers.) Brandr. is found in conifer woods and has rounder, smaller spores. Inedible.

Cortinarius rubellus
Cooke = *C. speciosissimus*

Cap 1–3in/2.5–7.5cm. The cap shape varies but is usually bluntly conical and the entire mushroom is a warm orange-brown to reddish-orange. The stem is usually rather spindle-shaped, rooting and the gills are widely spaced and broad. **Spores** rust-brown, 8–11 x 6.5–8.5μm, warty. In damp, mossy coniferous woods, especially spruce, recorded from Maine and possibly widespread in other northern states. Deadly poisonous, causes severe kidney damage.

Cortinarius pyriodorus
Peck

Cap 2–4in/5–10cm. The clear lavender colors, rather stout fruit-bodies with smooth cap, and stem with some whitish veil contrast with the cut flesh which is yellow-brown and often marbled. The odor is strong and penetrating of over-ripe pears. **Spores** rust-brown, 7–10 x 5–6μm, warty. Often common in conifer woods, widely distributed in North America. Inedible. This is sometimes considered a variety of the European *C. traganus* which is similar and smells of goat or acetylene gas.

Cortinarius sanguineus
(Wulf.) Fries
Cap 1–2in/2.5–5cm. The entire mushroom is bright blood-red to carmine-red, including the flesh. **Spores** rust-brown, 6–9 x 4–6µm, warty. In small groups in mossy conifer woods, widespread in North America. Inedible. The similar, rust-red species *C.marylandensis* is found in deciduous woods in the north-east.

Cortinarius semisanguineus
(Fr.) Gill.
Cap 1–2in/2.5–5cm. The yellow-brown, silky cap and stem contrast with the vivid cinnabar- to blood-red gills. The flesh is pale yellow to orange in the base. **Spores** rust-brown, 6–8 x 4–5µm, warty. Common in mixed woods in northern and eastern America. Inedible.

Hebeloma crustuliniforme
(Bull. ex St. Amans) Quel.
Cap 2–4in/5–10cm. The entire mushroom is the color of unbaked pastry, a dull ivory-buff. The gills, when mature, turn a pale gray-brown to tan and often have tiny beads of moisture on the edges. There is an odor or radish and the taste is slightly bitter. **Spores** dull cinnamon-brown, 9–13 x 5.5–7µm, minutely warty. Common in mixed woods and gardens, especially in the west. Poisonous.

Hebeloma edurum

Metrod ex Bon

Cap 2–6in/5–15cm. The dull, ochraceous cap is smooth with a slightly furrowed margin. The gills are pale buff and do not weep droplets. The stem is rather stout, club-shaped and fibrous-scaly with a slight ring-zone, browning at the base. The odor is fruity or like chocolate, then unpleasant. **Spores** dull cinnamon-brown, 9–12 x 5–7μm, almost smooth. Often in large numbers in conifer woods in the north-east. Poisonous.

Hebeloma syriense Corpse Finder

Karst.

Cap 1–2in/2.5–5cm. The brick-red cap is sticky and smooth, fading to ochre-brown. The gills are pale whitish-buff then cinnamon with minutely jagged edges. The stem is white to pale brown, slightly fibrous-scaly above. **Spores** pale cinnamon, 8–10.5 x 5–6μm, warty. This strange fungus grows on or near the remains of dead animals in woodlands in the north-east. Inedible.

Rozites caperata

(Pers.) Karst.

Cap 2–5in/5–12.5cm. The pale ochre cap is finely radially wrinkled and dusted with a fine white frosting of veil at the center. The gills are pale buff and join the stem. The stem is cylindrical, fibrous and has a white ring around the center and some white fragments of veil at the base. **Spores** rust-brown, 11–14 x 7–9μm, roughened. Frequent in conifer woods in the east, north and north-west of America. Edible.

Inocybe calamistrata
(Fr.) Gill.
Cap 1–2in/2.5–5cm. This dark brown species has a scaly cap and stem and the latter has a deep blue-green flush over the lower half. The odor is unpleasant, spermatic-fruity. **Spores** dull, earthy-brown, 9–12 x 4.5–6.5μm, elliptical. Common in conifer woods, often along stream edges, widely distributed in North America. Poisonous.

Inocybe tahquamenonensis
Smith
Cap ½–1in/1.2–2.5cm. This species is a contender for the darkest mushroom to be found. The deep purple-black to purple-brown cap and stem are both minutely scaly. The gills are nearly chocolate-brown with age. **Spores** 6–8 x 5–6μm, strongly angular, warted. An uncommon species in mixed woods in the north-east. Inedible.

Inocybe caesariata
(Fr.) Karst.
Cap 1–2in/2.5–5cm. The golden-yellow to tawny cap is rather flat and densely scaly. The gills are dull ochre and crowded and the stem is densely fibrous-scaly up to a ring-zone. **Spores** dull brown, 9–12.5 x 5–7μm, elliptical. Common in grass or leaf-litter along tracks and paths in woods, throughout the north-east. Inedible.

Inocybe pyriodora
(Pers. ex Fr.) Quel.
Cap 2–3in/5–7.5cm. The domed cap is densely fibrous and yellow-brown to pinkish-brown where stained. The gills are whitish to pale brown, crowded, and the stem is silky-fibrous to shaggy. The mushroom has a strong odor of ripe pears . **Spores** 7–10 x 5–7.5μm, elliptical. In deciduous woods, commonest in western states. Poisonous.

Inocybe geophylla
(Bull.) Karst.
Cap ½–1in/1.5–2.5cm. The domed cap is silky or smooth and varies from white to pale lilac (var. *lilacina*) as does the stem. The gills are pale brown and the mushroom has a strong, earthy or spermatic odor. **Spores** dull brown, 7–9 x 4–5.5μm, elliptical. Common in mixed woods throughout North America. Poisonous.

Gymnopilus spectabilis
(Fr.) Singer
Cap 3–6in/7.5–15cm. The bright golden-yellow to tawny-orange cap is coarsely fibrous to slightly scaly. Gills are crowded, shallow, orange to rust-brown and often speckled. Stem is large, tough and fibrous, often very swollen or club-shaped and has a well-developed membranous ring. It tastes extremely bitter. The flesh is yellowish and unchanging. **Spores** bright orange-brown, 8–10 x 5–6μm, coarsely warty. In clumps on dead deciduous wood. Scarce in north-east; often confused with *G. cerasinus* Peck, which is smooth, has more yellow gills, a fragile thin ring, a sweet, aromatic odor, bruises slightly greenish and is hallucinogenic; *G. spectabilis* is not.

Kuehneromyces mutabilis

(Schff.) Sing. & Smith = *Pholiota mutabilis*, = *Galerina mutabilis*
Cap 1–3in/2.5–7.5cm. The smooth cap is usually bicolored with a paler center as it dries out. The stem is scaly below the distinct ascendant ring. **Spores** rust-brown, 6–7 x 3–4.5μm, smooth. In clumps on dead wood in western states. Edible but must not be confused with ringed *Galerina* species (see below).

Galerina autumnalis

(Peck) Smith & Singer
Cap 1–2in/2.5–5cm. The flattened, sticky cap is dark brown to tawny when dry. The gills are attached, crowded and yellowish-rust. The stem is smooth, minutely lined and has a delicate ring above. **Spores** rust-brown, 8.5–10.5 x 5–6.5μm, roughened. In scattered troops (usually not clumped) on fallen logs throughout North America. Deadly poisonous.

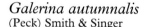

Phaeocollybia olivacea

Smith
Cap 2–4in/5–10cm. The bluntly conical olive-green cap is smooth, sticky to slimy. The gills are pale buff and join the stem. The stem is long and often deeply rooting, pale olive to reddish below. ***Spores*** rust-brown, 8–11 x 5–6μm, warty. An uncommon species, in coniferous woods, in western states.

STROPHARIA FAMILY
(BROWN-SPORED SECTION)

Most species grow on or near wood and may have glutinous, dry to scaly caps. The gills are sinuate-adnate and the spores are yellow-brown to rust and smooth. Usually a veil or ring is present.

Pholiota lenta
(Pers.) Singer

Cap 2–4in/5–10cm. The pallid, beige cap is very viscid with small white veil fragments near the edge. The gills are pale cinnamon to rust-brown and the stem is club-shaped with wooly scales. The odor is slight, of straw and the taste mild. **Spores** ochre-brown, 6–7 x 3–4µm. On the ground in leaf-litter on woody debris, usually under beech, in the north and west. Edibility uncertain.

Pholiota squarrosa
(Mull.) Kummer

Cap 2–6in/5–15cm. The dry, tawny-yellow cap is covered with recurved, pointed scales. The crowded gills are yellow to slightly olive, then rust-brown. The dry stem is also scaly up to the ring. **Spores** ochre-rust, 6–8 x 3–4µm. Often in large clumps on dead or dying deciduous trees, usually at the base, widely distributed in North America. Poisonous. The very similar *P. squarrosoides* differs in the cap being paler and distinctly sticky below the pointed, dry scales and has spores 4–6 x 2.5–3.5µm. Also common.

Pholiota flammans
(Fr.) Kumm.
Cap 1–3in/2.5–7.5cm. The brilliant
golden-yellow cap is both scaly and
sticky. The gills are bright yellow and
crowded. The stem is densely scaly up
to the ring-zone and is dry. **Spores**
ochre-brown, 4–5 x 2.5–3µm. Single
to small clusters on conifer logs and
stumps, widespread in North
America. Edibility doubtful.

Pholiota aurivella
(Fr.) Kumm.
Cap 2–6in/5–15cm. The deep orange-yellow cap is glutinous
when wet and has darker spot-like scales which can wash off.
The gills are crowded, pale yellow then tawny. The stem is
dry, yellowish, with recurved scales. **Spores** ochre-brown,
7–10 x 4.5–6µm. High up on standing trees, or on fallen logs,
usually deciduous, widespread. Edibility doubtful.

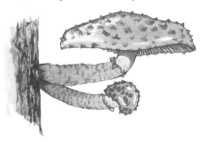

Pholiota highlandensis
(Peck) Smith & Hesler
Cap 1–2in/2.5–5cm. The sticky cap is
a dull reddish-buff to tawny-brown,
without scales. The gills are yellowish
then rust, slightly decurrent. The stem
is tawny and slightly fibrous up to the
whitish-yellow ring-zone. **Spores**
ochre-brown, 5–8 x 3.5–4.5µm.
Always on burn-sites in woodlands,
around burned stumps or logs,
widely distributed. Edibility
doubtful.

Pholiota malicola
(Kauff.) Smith
Cap 2–4in/5–10cm. Very smooth
for this genus, all parts of the
mushroom are yellow to tawny,
with only faint veil remnants at
the cap margin. There is a
fragrant odor, difficult to define.
Spores ochre-brown, 7.5–11 x
4.5–5.5µm. In clusters on logs
and stumps, widely distributed,
especially in the west. Edibility
unknown.

Pholiota albocrenulata
Peck
Cap 2–4in/5–10cm. Rich
reddish-brown to vinaceous-
brown, sticky with large
brownish-yellow veil
remnants. The gills are dark
grayish-brown with finely
toothed, white edges often
beaded with droplets. The
stem is reddish-brown with
darker scales below the ring-
zone. **Spores** brown, 10–15 x
5–8µm. Usually in ones or
twos on logs and living
deciduous trees, especially
maple and elm. Widely
distributed in North America.
Edibility doubtful.

BOLBITIUS FAMILY

These fungi have their cap cuticle made up of rounded instead of thread-like cells, and all have pale brown spores.

Bolbitius vitellinus
(Pers. ex Fr.) Fr.

Cap ¾–2in/2–5cm. The bright lemon-yellow cap is sticky or slimy and deeply grooved when old, and very fragile. The gills are bright cinnamon-brown and often liquefy in wet weather. The pale yellow-white stem is minutely hairy. **Spores** rust-ochre, 10–15 x 6–9μm, smooth. Found on dung, compost or old straw, common throughout North America. Edible but too fragile.

Agrocybe praecox
(Pers. ex Fr.)

Cap 1–3in/2.5–7.5cm. The tan-brown cap becomes paler when dry, is smooth and sometimes cracks in dry weather. The gills are attached and pale buffy-brown. The stem is slender and has a thin, membranous ring above. **Spores** dark brown, 8–11 x 5–6μm. Found in woods, flower beds in woodchip mulching, gardens, in spring and early summer, widely distributed. Inedible.

Agrocybe molesta
(Lasch) Singer = *A. dura*

Cap 2–4in/5–10cm. The whitish to pale tan cap soon cracks with age and dry weather. The gills are attached, crowded and pale buff then dark brown. The rather stout stem is stiff, smooth, whitish with a delicate ring which soon vanishes. **Spores** dark brown, 10–14 x 6.5–8μm, smooth. Often in groups, in grass or flower beds in mulch. Common in early summer in the north and north-east. Inedible.

Conocybe tenera
(Schaeff. ex Fr.)
Cap ³⁄₈–1in/1–2.5cm. Conical to bell-shaped, the cap is rich yellow-brown to reddish-brown, fading when dry. The thin gills are almost free, widely spaced and cinnamon-brown. The tall, thin stem is pale brown and very finely lined top to bottom. **Spores** reddish-brown, 11–12 x 6–7µm, smooth. In lawns, grassy edges of woodlands. Common, throughout North America. Inedible.

PINK SPORES

The pink-spored mushrooms have spores which range from pale pink to deep salmon, almost brownish-pink.

TRICHOLOMA FAMILY
(PINK-SPORED SECTION)

Although the majority of species in this family are white-spored, there are some exceptions, of which this is an example. Because the spores have a distinct pinkish hue, it is placed here in the guide, with other pink-spored genera, where one might expect to look first.

Rhodotus palmatus
(Bull. ex Fr.) Maire
Cap 1–3in/2.5–7.5cm. The beautiful reddish-pink to apricot cap has a thick, rubbery-gelatinous cuticle which is strangely wrinkled and pitted. The gills are attached to the stem and pale pink. The stout stem is pink and fibrous, rather tough, and usually set off-center. **Spores** pink, 6–8µm, globose and warty. On dead deciduous stumps and logs in the north and north-east. Inedible.

ENTOLOMA FAMILY

The *Entoloma* mushrooms and their relatives all have strange, angular pink spores, or long spores with angular ridges.

Clitopilus prunulus
(Scop. ex Fr.) Kummer

Cap 1–3in/2.5–7.5cm. The white to pale gray cap has a texture like kid-leather, and is often irregular and wavy in outline. The decurrent gills are white the pink. The short stem is smooth, white, often off-center. There is a strong odor and taste of fresh ground meal, bread dough or cucumber.
Spores pink, 10–12 x 5–7µm, with longitudinal ridges, appearing angular in end view. On soil in woods, throughout North America. Edible.

Entoloma velutinum
Hesler

Cap 1–3in/2.5–7.5cm. The blackish cap is minutely velvety and plush, soon deeply depressed and funnel-shaped. The gills are a beautiful pinkish-violet with dark violet-blue to violet-black, minutely toothed margins. The slender, fibrous stem is dark blue above shading to black below. The taste is cucumber-like. **Spores** pink, 8–11 x 6–8µm, 5-sided. Found in damp, mossy areas in mixed woods in the east and south-east, uncommon. Edibility unknown. There are a number of blue species of *Entoloma* which differ in shape of cap, colors of gills, and stem.

Entoloma abortivum
(Berk. & Curt.) Donk
Cap 2–4in/5–10cm. The slate-gray to pale gray cap is often finely streaked or silky, gills are broad, adnate to slightly decurrent and pinkish-buff. The stem is fibrous, whitish-gray and often swollen at the base. Odor and taste are cucumber-like. **Spores** pink, 8–10 x 5–6μm, 5–6-sided. Along with the normal fruit-bodies are found strange, swollen, brain-like bodies 1–3in/ 2.5–7.5cm across, which are white, soft, with pink masses of tissue inside. These aborted forms are the result of attack by *Armillaria mellea*. Common and widespread in deciduous woods through much of North America. Considered a delicacy in the aborted form. The normal form is best avoided as it is easy to mistake for other species.

Entoloma salmoneum
(Pk.) Sacc.
Cap ³⁄₈–1⁵⁄₈in/1–4cm. The bell-shaped or sharply conical cap is a lovely salmon-orange, paler with age. The gills are salmon-pink. The tall, slender stem is fibrous, pale salmon. **Spores** pink, 10–12 x 10–12μm, 4-sided. In leaf-litter near conifers, in the north, east and south-east, common. Inedible.

Entoloma murraii
(Berk. & Curt.) Sacc.
Cap ³⁄₈–1¹⁄₄in/1–3cm. The cap is sharply conical with a point or pimple at the center, bright yellow to yellow-orange, silky. The gills are yellow then pinkish. The tall stem is fibrous, yellow. **Spores** 9–12 x 8–10μm, 4-sided. On the ground in damp, swampy woodlands, common from the north, southward through the east to Alabama. Inedible.

Entoloma luteum
Peck

Cap ⅜–1in/1–2.5cm. Bluntly conical the cap is a greenish-yellow to yellow-buff, silky to slightly roughened at the center. The gills are yellowish then pink. The tall stem is fibrous, whitish above, greenish-yellow below. **Spores** pink, 9–13 x 8–12µm, 4-sided. On mossy ground in mixed woods, uncommon, from the north through the east and south-east. Inedible.

Entoloma lividum
Quelet

Cap 2–6in/5–15cm. A large, fleshy species, the cap is ivory-white to pale buff, smooth, often slightly streaky or mottled. The gills are sinuate, yellow-ochre when young then pink when mature. The stout stem is fibrous, whitish. The odor and taste are strong of fresh meal or flour. **Spores** pink, 8–10 x 7–9µm, 5–7-sided. In grassy woodlands on calcareous soils, north-eastern North America. Poisonous, causes severe gastric upsets, can be deadly.

Entoloma strictius
(Peck) Sacc.

Cap 1–3in/2.5–7.5cm. The domed cap is a pale gray-brown, paler when dry, smooth. The gills are attached to nearly free, broad, pink. The long, straight stem is silky-fibrous, whitish-gray and often twisted, the base has a wooly covering of mycelium. **Spores** pink, 10–13 x 7.5–9µm, 5–6-sided. Common in leaf-litter or on rotted wood, in wet northern areas southward through the east to Florida. Poisonous. A number of similar species can only be separated microscopically.

PLUTEUS FAMILY

These all have gills completely free of the stem and deep salmon-pink spores. One genus (*Volvariella*) has a volva at the base of the stem.

Pluteus cervinus
(Schaeff. ex Fr.) Kumm.
Cap 2–4in/5–10cm. The dull gray-brown to dark brown cap is smooth but with fine fibers and occasionally tiny scales at the center. The gills are free of the stem and start white, then mature pink. The fibrous stem is white above shading to brown below. **Spores** pink, 5.5–7 x 4–5μm, smooth. Very common mushroom growing on dead wood, sawdust piles and buried wood throughout North America. Edible but tasteless. On the edges of the gills are bottle-shaped cystidia with 2–3 hook-like projections.

Pluteus petasatus
(Fr.) Gill.
Cap 2–6in/5–15cm. This almost white species has a faint flush of brown in its cap, which is fibrous-scaly at the center. The gills are often very deep and rounded at the margin, and usually stay pale for a long time before turning pink when mature. The stout stem is fibrous, slightly brown below. **Spores** pink, 6–10 x 4–6μm. Common on old deciduous stumps, sawdust or woodchips; widely distributed. Edible and good.

Pluteus romellii
(Britz.) Sacc.
Cap 1–2in/2.5–5cm. The deep brown, rather wrinkled or veined cap contrasts with the clear yellow of the stem. The young gills are also pale yellow before turning pink. **Spores** pink, 6–7 x 5–6μm, smooth. Quite a common species on old logs of deciduous trees, widespread throughout North America. Edibility uncertain.

Pluteus aurantiorugosus
(Trog.) Sacc.
Cap 1–2in/2.5–5cm. The brilliant scarlet cap fades to orange as it expands, while the gills when young are yellow. The stem is whitish above but orange-yellow below. **Spores** pink, 5.5–6.5 x 4–4.5μm. This small, but spectacular species occurs on dead elm and maple but is sadly not common. Edible.

Volvariella bombycina
(Schaeff. ex Fr.) Singer
Cap 2–6in/5–15cm. This magnificent mushroom has a silky-shaggy cap of white tinged with yellow. The broad, free gills are pink and the white stem emerges from a large, egg-like volva. **Spores** pink, 6.5–10.5 x 4.5–6.5μm. It grows out of holes, or wounds in the trunks of elms, maples and beech trees and sometimes on old stumps. It is a widespread but rather uncommon species. Edible and good.

Volvariella speciosa
(Pers. ex Fr.) Singer
Cap 2–4in/5–10cm. The smooth, white to grayish cap is often very sticky or viscid. The gills are free, broad, and pink when mature. The often long stem is white and comes from a thick, deep volval cup. **Spores** pink, 11.5–21 x 7–12μm. Quite common in some years in fields in old grass, rotting straw or around stables. Edible, but because it grows on the ground, and has a volva, one might easily pick a poisonous *Amanita* instead. It is safer to buy the canned Paddy-straw Mushrooms (*V. volvacea*).

GASTEROMYCETES

PHALLALES, STINKHORNS AND SQUID FUNGI
PUFFBALLS AND EARTHBALLS
BIRDS NEST FUNGI

Although this is now considered by most mycologists to be a rather artificial grouping of fungi which are not really closely related, it is nevertheless a very convenient one for use in guide books, since it brings together all those fungi which produce their spores inside the fruit-body, rather than on an external hymenium. The Gasteromycetes rely to a great extent on external forces – wind, rain, insects – to carry off their spores and have evolved some of the strangest forms to accomplish this.

PHALLALES, STINKHORNS AND SQUID FUNGI

All species emerge from an "egg" and expand rapidly in a matter of hours to full size. Their spore-mass (gleba) liquifies and gives off an unpleasant odor to attract insects which eat the spores and unknowingly pass them on to germinate elsewhere. All the stinkhorns can be "hatched" by placing large, unopen eggs on a damp paper towel under a glass.

Phallus hadriani

Vent.
4–6in/10–15cm high. The "egg" is tinted pinkish-mauve and the spongy stem is white. The thimble-like cap is pitted and ridged with a large disc-like opening at the apex. Covering the cap is the dark green spore mass which liquifies and has an odor of rotting flesh or vegetables. **Spores** 4–5 x 2µm. Grows alone or in groups in gardens, lawns, shrubberies, in rich, sandy soils, especially common in the west and central states. Edible before "hatching" but best avoided. This is often called *P. impudicus*, the famous Stinkhorn of Europe, but that species is not reliably recorded yet from North America; it differs in having a white egg, stouter fruit-body with smaller aperture at the top and is always associated with dead wood.

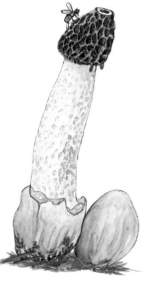

Phallus ravenellii
Berk. & Curt.

4–6in/10–15cm high. The cap of this species is minutely granular, roughened (not ridged or pitted like *P. hadriani*) under the green spore mass. The spongy yellowish-white stem emerges from a pinkish egg. **Spores** 3–4.5 x 1–2µm. Grows on rotted stumps, sawdust, often in clusters, southward through eastern states to Florida, especially in disturbed, urban areas. Edible when young.

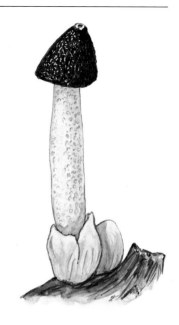

Dictyophora duplicata
(Bosc.) Fisch.

4–6in/10–15cm high. This remarkable species has a pitted-ridged cap from underneath which a white, net-like skirt hangs down. The white, spongy stem emerges from a large white egg. The spore-mass is green, and soon liquifies to give an unpleasant, fetid odor. **Spores** 3.5–4.5 x 1–2µm. Quite common around old stumps and trees, particularly in the east and south-east.

Mutinus caninus   Dog Stinkhorn
(Pers.) Fr.

2–5in/5–12cm high. This is the most often misidentified species of *Mutinus* in North America. The true *M. caninus* is rather uncommon and appears to be distinctly northern in distribution. It is distinguished by its long, slender stem with a distinct head which is usually pinched in and narrower than the stem. The head is orange while the stem is usually white. The cells of the stem form open chambers (in contrast to *M. elegans*) and the eggs are white. **Spores** 4–5 x 1.5–2μm. This is often confused with *M. ravenelii*. The spore mass has a very weak odor and may go almost unnoticed. It grows in leaflitter in woodlands.

Mutinus ravenelii

(B.& C.) Fischer

2–4in/5–10cm high. This
is a common species in
eastern and central states
down through Florida but
is usually misidentified as
M. caninus. It is easily
distinguished by its
bright, rosy-red
coloration of head and
stem; shorter, stouter
fruit-body; swollen head
which is thicker than the
stem, and very strong,
foul odor easily detectable
from a distance. **Spores**
4–5 x 1.5–2μm. It grows
in mixed woods although
often in coniferous woods
(unlike *M. caninus*) in
leaf-litter and woody debris.

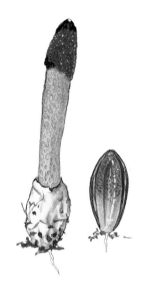

Mutinus elegans Elegant Stinkhorn

(Mont.) Fisch.

4–7in/10–18cm high.
Common in the north-
east, this species is
becoming even more
abundant through the
expanding practice of
using woodchip mulches
on flower beds. The stem
tapers uniformly to an
elegant point *without* a
well-defined head. The
cells of this stem are
globose and do not form
open chambers. **Spores**
4–7 x 2–3μm, spread over
the upper half of the
stem. The eggs are often
pinkish-purple. The odor
is fetid but almost sweet
or metallic, not too
unpleasant.

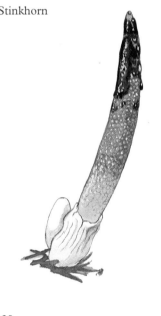

Pseudocolus fusiformis Stinky Squid
(E. Fisch.) Lloyd

2–4in/5–10cm high. The common name is very apt, as it does look remarkably squid-like. The pale orange arms vary from 3 to 4 in number although 3 is most usual. They are joined at the tips when young but often separate when older. **Spores** 4.5–5 x 2–2.5µm. The spore mass lines the inner surface of the arms and is strongly fetid. The small eggs are white. This species was probably introduced from Southeast Asia but is spreading throughout the north-east on woodchip garden mulches and occasionally in woodlands.

Clathrus ruber Cage or Lattice Stinkhorn
Mich. ex Pers.

2–6in/5–16cm high. Once again, an unmistakable fungus with its thick, orange-red, cage-like structure. Spores line the inner surface of the cage and are strongly fetid. The egg is white, often with mycelial cords at the base. **Spores** 5–6 x 1.5–2.5µm. It is found, sometimes in large numbers, on woodchips, humus, usually in gardens or other landscaped areas, scattered across North America from Florida to California but is rarer further north.

PUFFBALLS AND EARTHBALLS

As their common name suggests, they are ball-like and the spores which are inside the ball become dry and powdery, and will puff out if the ball is tapped.

Lycoperdon pyriforme
Schaeff. ex Persoon

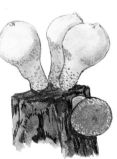

⅝–¾in/1.5–4.5cm across. These are rather pear-shaped puffballs and start white before soon turning yellowish-buff. The skin is rather smooth. At the bottom of the very short sterile base are white strands running into the rotten wood on which it grows. **Spores** olive-brown, 3–4.5µm, globose and smooth. Often grows in large clusters on dead wood, common and widely distributed. Edible when young and white.

Lycoperdon perlatum
Persoon

1–2in/2.5–5cm across. The white, rounded to slightly club-shaped ball has small white spines or warts, often in tiny rings which are easily rubbed off leaving round marks. A short to rather distinct sterile base is present. The spore-mass is white then greenish-ochre when mature. **Spores** 3.5–4.5µm, globose and minutely warted. Common in fields, roadsides and gardens, widely distributed. Edible when young and if white and completely uniform in appearance.

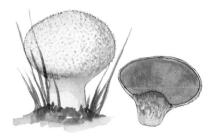

Lycoperdon foetidum

Bonord.

1–2in/2.5–5cm across. This is the only puffball to have tiny, dark brown spines, even when young. The whole puffball matures to a dark yellowish hue, contrasting with the blackish spines. The spore-mass is dull sepia-brown. **Spores** 4–5µm, globose and minutely spiny. Common in woods, especially conifers in the western states. Edible when young, it smells rather unpleasant when old, as do many other species.

Calvatia sculpta

(Hark.) Lloyd

2–6in/5–15cm across. A remarkable species with thick, sculptured pyramid-like warts or spines which crack apart and fall away with age exposing the spore-mass beneath. **Spores** olive-brown, 3.5–6.5µm, globose and minutely spiny. Fairly common in the mountains of the west, often under conifers. Edible when young.

Calvatia cyathiformis
(Bosc.) Morgan

3–6in/7.5–15cm across. A large, balloon-like to club-shaped species with a prominent sterile base, the ball becomes minutely cracked with age and eventually falls away to leave the sterile, cup-shaped base behind. The spore-mass is a deep purple-brown. **Spores** 3.5–7.5µm, round with distinct spines. Common in fields roadsides, golf courses, in eastern and central states. Edible when young and white within.

Calvatia craniformis
(Schw.) Fries

3–8in/7.5–20cm across. Often huge, this top-shaped, swollen species soon becomes puckered, brain-like, and the skin flakes away in plates to reveal the greenish-yellow spore-mass. **Spores** 2.5–4.5µm, globose and almost smooth. Common on open ground near oaks, widely distributed. Edible when young. *C. bovista*, also large and top-shaped, with olive-yellow spores, is less wrinkled and brain-like and has spores 4–6.5µm across.

Langermannia gigantea Giant Puffball

(Batsch.) Rotsk. = *Calvatia gigantea*

8–20in/20–50cm across. This often huge species (specimens as large as a small sheep have been found) forms a smooth white ball or flattened oval with a skin the texture of kid-leather. As it matures this flakes away to expose the yellow-brown spore-mass. **Spores** 3.5–5.5μm, globose, minutely warted. Quite common in some years in hedgerows, fields and gardens in eastern North America west to Ohio. Edible and good when young and white throughout. *L. booniana* is found in western states. It differs in its thick skin being covered in large flattened warts and cracks.

Geastrum triplex Earthstar

Jung.

1–3in/2.5–7.5cm across. The leathery, gray-brown, onion-like ball splits open to form a star with a pinkish-buff, smooth to cracked surface. At the center is a ball containing the spore-mass, this is set in a distinct, cup-like collar. **Spores** dark brown, 3.5–4.5μm, globose, warted. In leaf-litter in woods, especially beech, widely distributed. Inedible.

Geastrum pectinatum
Persoon

1–2in/2.5–5cm across. The gray-brown, tough ball splits to form a star. At the center is a rounded gray-white ball set on a slender stalk. The ball has a narrow, beaked, deeply grooved opening on the top where the spores are released. **Spores** brown, 4–6μm, globose, warted. On the ground in woods and gardens, especially near cedars, eastern North America. Inedible.

Scleroderma citrinum Common Earthball
Persoon

2–4in/5–10cm across. The earthballs differ from the true puffballs in their thick, leathery skins and spores which do not all mature simultaneously. This species has a yellow-ochre, warty-scaly thick skin and a spore-mass which starts white but soon becomes marbled with purple-black. The skin flakes away slowly over many weeks to release the spores. **Spores** 8–12μm, with spines and a fine network. The mature fungus has a strong, pungent odor difficult to define. It grows on soil in woodlands throughout much of North America. Poisonous.

Scleroderma polyrhizon
(Gmelin) Persoon

2–6in/5–15cm across. This large earthball begins life completely buried in sandy soil and slowly forces its way up through the surface. The thick, pale yellow skin then splits and peels back to form a star-like structure, and reveals the dark brown, powdery ball inside which is the spore mass. **Spores** 5–10μm, warted. Common in sandy soils, widely distributed. Probably poisonous.

Astraeus hygrometricus
(Pers.) Morgan

1–2in/2.5–5cm across. Starting as a dark brown, onion-like ball in the soil it splits to form a star, exposing a rounded, gray-white ball at the center. The spore-mass in the ball is brown. **Spores** 7–11μm, warted. The arms of the "star" become cracked as they bend back and will close back up if the weather turns dry, only to reopen when it becomes moist again. Common in sandy soils throughout North America. In the west is found *A. pteridis*, which is larger and with a very cracked, quilt-like pattern on the arms.

Pisolithus tinctorius Dye–makers Mushroom
(Pers.) Coker & Couch

2–4in/5–10cm across, 2–8in/5–20cm high. A rather ugly
fungus, it is often buried deep in the soil. It forms an irregular
club-shaped structure, ochre to reddish-brown with a variable
length of stem. The spore-mass forms a mass of whitish
to yellow eggs embedded in a
blackish jelly, eventually
forming a brownish powder.
This crumbles away leaving
the sterile base in the soil.
Spores 7–12μm, spiny.
Common in sandy soils in
open pine woods, widely
distributed. Inedible. The
fungus is boiled down to
make a rich golden-brown
to black dye.

Tulostoma brumale
Persoon

Head about ⅝in/1.5cm across, **stem** about 1½in/3cm
high. Like a small puffball on a slender, bulbous
stem, this species grows in dry, sandy soils. **Spores**
brown, 3–5μm, globose, minutely warted. Widely
distributed in North America. Inedible. There are
a number of other very similar species, differing in
minor details.

Calostoma cinnabarina
Desv.

1½–2in/4–5cm high. This
extraordinary fungus is rather like a
small, bright red tomato set on a
spongy stem. The entire fungus is
enclosed in a thick gelatinous veil at
first and this breaks away with pieces
of the red inner skin and remains
as small "pips" around the base.
There is a small, deep red star-
like mouth at the top of the ball
where the spores exit. **Spores**
pale yellow, 14–20 x 6–8.5 μm, distinctly pitted. Frequent in
mossy banks and tracks, especially by streams, throughout
eastern states south to Florida. Edibility uncertain.

BIRDS NEST FUNGI

As the common name suggests, these fungi look like tiny birds' nests with minute eggs inside. The eggs are in fact the spore-masses.

Cyathus striatus Birds-Nest Fungus
(Huds.) Willd.

¼–⅝in/1–1.5cm across. The urn-shaped structure is hairy on the outside with a smooth, gray, fluted interior. At the bottom are tiny dark eggs attached by coiled cords. These "eggs" are the spore-masses and are thrown out of the cup by splashing raindrops. **Spores** 15–20 x 8–12μm, smooth but notched at one end. Common on dead twigs, wood-chips throughout North America.

Cyathus olla Common Birds-Nest
(Batsch.) Persoon

¼–⅜in/0.5–1cm across. This funnel-shaped species is brown and roughened on the outside, but smooth and white on the inside. The "eggs" are grayish. **Spores** 11–13 x 7–8μm, smooth. Common in fields, gardens, attached to woody debris, widely distributed.

Crucibulum laeve

(Huds.) Kamb.
¼–⅜in/0.5–1cm across. The almost
cylindrical nest is tawny-yellow with a
hairy lid covering it at first. This splits
open to reveal the smooth, inner
surface. The "eggs" are whitish.
Spores 4–10 x 4–6μm, smooth. On
dead wood and woody debris, widely
distributed. The only birds-nest
fungus with white eggs.

Sphaerobolus stellatus

Tode ex Persoon
¹⁄₁₆in/1.5mm across. This tiny fungus
forms a star-like structure with a
single, brown "egg" at the center. The
egg is catapulted out by the sudden
flipping outward of the inner wall of
the star. The egg can be thrown
several feet. **Spores** 7.5–10 x
3.5–5μm, smooth. Common but
easily overlooked on rotten wood,
twigs, and horse dung, in eastern
states.

CHANTERELLES

These often edible fungi include the well-known Chanterelle eaten all over the world. They all lack true gills, forming their spores on either a smooth undersurface or on blunt wrinkles or ridges.

Craterellus fallax Black Trumpet
A.H. Smith

1–3in/2.5–7.5cm across. These black, funnel-shaped fungi are very thin and completely hollow down the center. The inner surface may be slightly scaly and becomes gray when dry. The outer surface becomes flushed with pinkish-buff as the spores mature. **Spores** pinkish-buff, 10–20 x 7–11μm, smooth. Common and often in large numbers on damp, mossy banks in deciduous woods. Edible and good. The very similar *C. cornucopiodes* has white spores.

Cantharellus cibarius Chanterelle
Fries

1–6in/2.5–15cm across. The yellow-orange cap is often wavy and irregular, with an inrolled margin, and is very fleshy. The undersurface is formed of numerous ridges and wrinkles, often cross-veined and descending the short, stout stem. The flesh is white. The odor is pleasant of apricots. **Spores** pale buff, 8–11 x 4–6μm, smooth. In groups in mixed woods, especially oak and pine, common throughout North America. Edible and delicious. *C. lateritius*, commoner in the north-east, differs in its completely smooth undersurface, but is otherwise similar.

Cantharellus tubaeformis
Fries

1–3in/2.5–7.5cm across. The yellow-brown cap is thin and often depressed at the center. The blunt wrinkles and ridges on the underside are yellow then grayish-violet and run down the stem. The stem is slender, hollow and yellow-orange. **Spores** cream, 8–12 x 6–10µm, smooth. Common in sphagnum moss in boggy woods in eastern and northern North America. Edible and good.

Cantharellus cinnabarinus Cinnabar Chanterelle
Schw.

³⁄₈–2in/1–5cm across. This small chanterelle is a bright cinnabar-red fading to pinkish-red with age. The undersurface is wrinkled and ridged. **Spores** pinkish-cream, 6–11 x 4–6µm, smooth. Often in large drifts on mossy banks of streams and paths in mixed woods, common throughout the northern and eastern states down to Florida. Edible and good.

Cantharellus cinereus
Persoon

1–3in/2.5–7.5cm across. This thin-fleshed, gray-black mushroom is funnel-shaped, rather like *Craterellus fallax* also shown here, but differs in having well-developed wrinkled "gills" on the undersurface. **Spores** white, 8–10 x 5–7µm, smooth. Uncommon, on damp mossy banks in deciduous woods in the north-east. The form usually found is the variety *multiplex* which is often many trumpets fused together. Edible.

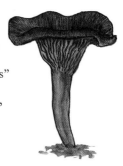

Cantharellus ignicolor
Pet.

³⁄₈–2in/1–5cm across. A bright, yellow-orange species, it is thin-fleshed with a sunken center to the cap. The undersurface has well-developed ridges and wrinkles running down the slender stem, which is hollow with age. **Spores** pale salmon-ochre, 9–13 x 6–9μm, smooth. Quite common in mixed woods in the northern and eastern states. Edible.

Cantharellus minor
Peck

¼–1¼in/0.5–3cm across. This often tiny chanterelle is bright yellow-orange and thin-fleshed. The undersurface has a few, narrow, forked ridges, widely spaced. **Spores** pale yellow-orange, 6–11.5 x 4–6.5μm, smooth. Common on bare soil or moss by trackways and streams in deciduous woods. Edible.

Cantharellus subalbidus
A.H. Smith & Morse

2–6in/5–15cm across. A whitish chanterelle, very thick-fleshed with an inrolled margin to the cap. The undersurface is ridged and wrinkled and descends the short, stout stem. The flesh is solid, white. All parts bruise rust-yellow to brown. **Spores** white, 7–9 x 5–6μm, smooth. In mixed woods in the Pacific Northwest, quite common. Edible and good.

Gomphus floccosus
(Schw.) Sing.

Cap 2–6in/5–15cm. Looking just like a tall scaly vase, the yellow-orange, funnel-shaped upper surface has coarse, recurved scales. The outer surface is cream to ochre, wrinkled and ridged, running down the short stem. **Spores** ochre, 11.5–14 x 7–8μm, warty. In mixed woods throughout North America. Inedible.

Polyozellus multiplex
(Under.) Murr.

1–4in/2.5–10cm across. This remarkable fungus, often found in large clusters fused together, is a deep blue to purplish-black. Each cap is irregular, spoon-shaped or funnel-like, smooth on top with a wrinkled undersurface that is purplish with a white bloom like a grape. The odor is aromatic. **Spores** white, 6–8.5 x 5.5–8μm, angular and warted. On the ground under spruce and fir, northern North America, west to the Rockies. Edible.

CORAL & CLUB FUNGI

As their common name suggests, these fungi resemble undersea corals or colored clubs; many are among the brightest colored of fungi and can reach large sizes. Their spores are produced on basidia on the outer surface of each club.

Ramaria stricta

(Fr.) Quel.
1–4in/2.5–10cm across. This yellow-brown coral has many straight, parallel branches held very upright. The stems fuse below into a central stalk, the flesh is brownish-white. All parts bruise darker on handling. The odor may be unpleasant and the taste bitter. **Spores** golden-yellow, 7–10 x 3.5–5.5µm, minutely warted. On deciduous or coniferous wood throughout North America, frequent. Inedible.

Ramaria botrytis

(Fr.) Rick.
2–6in/5–15cm across. A beautiful, cauliflower-like coral with many densely branched clubs, whitish with pinkish-purple tips. The flesh is white. **Spores** pale ochre, 11–17 x 4–6µm, with longitudinal lines. Under spruce and fir, widely distributed. Edible but easily confused with other less edible species, i.e. *R. formosa* has pinkish branches with yellowish tips.

Clavicorona pyxidata
(Fr.) Doty

2–4in/5–10cm across. A common species on fallen logs this dull ochre-yellow species has numerous forking branches arising from a single stem. Each branch is tipped with a tiny cup-like crown with minute points. The taste is peppery. **Spores** white, 4–5 x 2–3μm, smooth. On dead willow, poplar and aspen throughout North America. Edible.

Clavariadelphus pistillaris
(Fr.) Donk

3–8in/7.5–20cm high, 1–2in/2.5–5cm across. Forming a single, large club, the color ranges from pale ochre to pinkish-brown and it bruises darker brown. The flesh is white and bitter to taste. **Spores** creamy-white, 11–16 x 6–10μm, smooth. In leaf-litter in deciduous woods, especially beech, eastern North America and on the west coast under oak. Edible but poor.

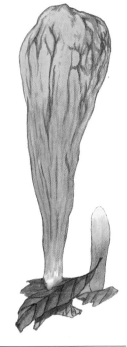

Clavariadelphus truncatus
(Quel.) Donk
2–6in/5–15cm high,
1–3in/2.5–7.5cm across.
Forming a flat-topped,
wrinkled club from
pinkish-brown to yellow-
ochre in color. The flesh is
white and mild to the taste.
Spores ochre, 9–13 x
5–8μm, smooth. Under
conifers throughout North
America. Edible. *C.
borealis*, similar but with
more lilac tints and a white
spore print, is found in the
Pacific Northwest.

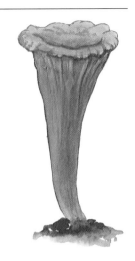

Sparassis crispa Western Cauliflower Mushroom
Wulf. ex Fr. = *S. radicata* 6–12in/15–30cm across. Looking
very much like a large cauliflower with numerous flattened,
twisted and crisped, leaf-like branches, pale cream-yellow in
color. The odor is slightly spicy. It often has a rooting base
buried deep in the ground. **Spores** white, 5–7 x 3–5μm. At the
base of conifers, common in western North America. Edible
and delicious. The Eastern Cauliflower Mushroom,
S. herbstii, differs in its larger,
looser and more flattened
branches and grows at
the base of oaks,
equally edible.

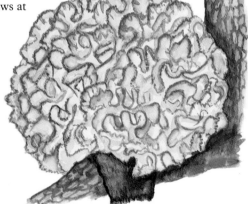

Clavulinopsis fusiformis

(Sow.) Corner
2–6in/5–15cm high. This
golden-yellow fungus
forms a clump of simple,
pointed clubs, hollow
inside. **Spores** creamy,
5–9 x 4.5–9μm, smooth.
Common in grassland
and open woods, widely
distributed. Edible.

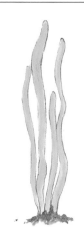

Clavaria vermicularis

Fries
2–6in/5–15cm high. The
Latin name means worm-like
and this does suggest a
cluster of white, wavy
worms. The clubs are hollow
and very brittle. **Spores**
white, 5–7 x 3–4μm, smooth.
In grassy places and open
woodlands, widely
distributed. Edible. The
similar species *C. fumosa*
differs in its grayish-lilac
clubs.

Clavulina cristata

(Fr.) Schroet.
1–3in/2.5–7.5cm across and high. A densely branched, coral-like species with a central stem and finely pointed tips, it varies from pure white to yellowish. **Spores** white, 7–11 x 6.5–10µm, nearly round and smooth. In mixed woodlands throughout North America Edible. The similar *C. cinerea* has blunter, grayish-white branches.

Clavulina amethystina

(Fr.) Donk
1–4in/2.5–10cm across and high. A beautiful lilac-violet coral-like species with many branched clubs. **Spores** white, 7–12 x 6–8µm, smooth and growing on 2-spored basidia. In deciduous woods in eastern North America, frequent. Edible. The very similar *C. zollingeri* differs in only slightly branching and having 4-spored basidia with smaller spores.

Thelephora terrestris

Fr.
1–3in/2.5–7.5cm across. Forming a cluster of overlapping, fan-shaped or vase-shaped caps, dark, blackish-brown and hairy-fibrous. The margin in often whitish and torn. The underside is smooth to wrinkled or minutely lumpy. There is usually an earthy, moldy odor. **Spores** purplish-brown, 8–12 x 6–9µm, minutely spiny. Common on soil in mixed woods, widely distributed. Inedible.

CRUST FUNGI

These form crust-like sheets on wood and occasionally spread out into small shelves or brackets. They do not have tubes on the undersurface like the true bracket fungi (Polypores).

Chondrostereum purpureum Silver Leaf Fungus
(Fr.) Pouz.

4–6in/10–15cm across or larger. This fungus forms irregular, leathery sheets with fused, overlapping caps and is colored a lovely bright lavender-purple on the edge and undersurface, a dull yellow-ochre above. **Spores** white, 5–6.5 x 2–3μm, smooth. On apple, plum and other fruit and deciduous forest trees, common. This fungus attacks the tree which shows a characteristic silvery blight on the leaves, and finally sickens and dies.

Stereum hirsutum
(Willd. ex Fr.) S.F. Gray

¼–1in/0.5–2.5cm across. Many of the flattened, shell-like caps may be fused together in rows, each one is concentrically zoned in shades of brown, cream or gray and the upper surface is minutely hairy. The lower surface is smooth, yellow-gray. **Spores** white, 5–8 x 2–3.5μm, smooth. On dead deciduous wood, especially birch and beech, widely distributed. Inedible.

TOOTHED FUNGI

These fungi all form their spores on projecting conical teeth which usually hang down from the undersurface of the cap or fruit-body. Some species grow on the ground and have a stem, others form shelf-like brackets on trees.

Hydnellum spongiosipes

(Pk.) Pouz.

Cap 2–4in/5–10cm. The very tough, almost woody cap is convex to flattened and rich reddish-brown, minutely hairy. The tiny, densely crowded spines on the undersurface are dark brown and run down the short, very bulbous, velvety stem. Several caps may be fused together. The flesh is two-layered with a dark, zoned, inner core. **Spores** brown, 5.5–7 x 5–6μm, warted. Common, in groups in leaf-litter under oaks, widely distributed. Inedible. There are numerous other species, many with strong tastes and odors and colored flesh, i.e. *H. caeruleum* with flesh banded with blue and orange-brown stem flesh, and mealy odor.

Sarcodon imbricatum
(L.) Karst.

Cap 4–6in/10–15cm. The gray-brown cap is coarsely scaly and the undersurface has densely packed, tiny gray-brown teeth which run down the short stem. The flesh is soft, pale brown and may taste a little sharp. **Spores** brown, 6–8 x 5–7µm, with large warts. Common under mixed woods throughout North America. Edible. Other species include *S. scabrosum* with blackish-green stem base and very bitter taste, and *S. joeides* with pale lavender flesh which turns green with KOH solution.

Hydnum repandum
Linnaeus = *Dentinum repandum*

Cap 2–4in/5–10cm. The smooth, irregularly shaped cap is a pale orange-buff, the teeth are white to pale pinkish-yellow and descend the short, whitish-yellow stem. The flesh is white bruising orange-brown and has a mild to slightly bitter taste. **Spores** white, 6.5–9 x 6.5–8µm, nearly round, smooth. Common, often in large numbers in leaf-litter under mixed trees, throughout North America. Edible and good. Two similar but white species found in the south are H. *albomagnum* which is mild tasting and does not stain when bruised, and *H. albidum* which is peppery and bruises yellow-orange.

Hydnum umbilicatum
Peck

Cap ½–1in/1.3–2.5cm. A small, reddish-brown species with the center of the cap open, forming a hole down into the stem. The spines are pale orange-buff. **Spores** white, 7.5–9 x 6–7µm, smooth. Common under conifers in the north-east. Edible.

Hericium ramosum
(Bull. ex Mer.) Let.

4–10in/10–25cm across. This beautiful, large, but delicate species is like an undersea coral growing on a tree. Composed of a mass of narrow arms which repeatedly branch and have short spines all along the lower surfaces. **Spores** 3–5 x 3–4µm, white, finely roughened. Frequent on old deciduous logs throughout North America. Edible and delicious. The similar *H. americanum* (=*H. coralloides* in part) has much longer spines grouped at the ends of each branch only, equally edible. *H. abietis* from the Pacific Northwest is pinkish-buff, also with spines grouped at the branch ends.

Hericium erinaceus
(Fr.) Pers.

4–10in/10–25cm across. Forming a
compact oval ball, all the spines are in
a dense mass and are quite long,
about 1–2in/2.5–5cm. **Spores** white,
5–6.5 x 4–5.5μm, minutely
roughened. This attractive
white fungus grows,
often very high up, on
living trees of beech,
oak and maple, widely
distributed. Edible and
delicious when young.

Auriscalpium vulgare Earpick Fungus
S.F. Gray

Cap ⅜–1in/1–2.5cm. The cap of this
small species is dark brown, minutely
hairy and with the slender, hairy stem
attached to one edge. The tiny spines
are whitish-brown. **Spores** white, 5–6
x 4–5μm, minutely spiny. Found only
on fallen, rotting pine-cones this
small, dark species is often
overlooked. Widespread in northern
North America. Inedible. The generic
name is Latin for earpick, an
instrument used by the Romans for
personal hygiene.

POLYPORE FUNGI, BRACKET AND SHELF FUNGI

All fleshy or woody fungi with their spores produced inside tubes (except for the Boletes) are included here. Many are very large and woody and are common on our woodland trees. Some grow on the ground and can look like a bolete but are much tougher fleshed. A number are serious parasites of trees.

Polyporus brumalis Winter Polypore
Fries

Cap 1–4in/2.5–10cm. One of a number of polypores with a cap and more or less central stem, this species has a smooth, yellow-brown cap, whitish tubes with pores spaced 2–3 per mm, and a slender gray-brown, minutely hairy stem. **Spores** white, 5–7 x 2–2.5μm, sausage-shaped, smooth. On fallen timber, especially birch from late fall throughout the winter and into spring, widespread. Inedible.

Polyporus varius
Fries = *P. elegans*
Cap 2–4in/5–10cm. The cap is
pale cinnamon-buff, smooth,
and the tubes and pores are
yellowish to pale brown, 4–5
per mm. The almost central
stem is slender, pale tan with a
black lower half. **Spores** white,
7–10 x 2–3.5μm, sausage-
shaped, smooth. On fallen twigs
and logs of deciduous wood
throughout North America.
Inedible. *P. badius* has a larger,
reddish-brown, shiny cap and
often completely black stem.

Polyporus squamosus   Dryad's Saddle
Fries
Cap 4–12in/10–30cm. A magnificent species looking rather
like a seat or saddle growing out of dead or standing trees. The
cinnamon-brown cap has broad, darker brown, flattened
scales, and the pale cream-buff pores are large, honeycomb-
like and descend the short, tough stem. The latter has a black
base. **Spores** white, 10–16 x 4–6μm, cylindrical, smooth.
Widespread in North America. Edible and good when very
young.

Grifola frondosa Hen of the Woods
(Fr.) S.F. Gray

Individual caps 1–3in/2.5–7.5cm, entire **fruit-body** 6–20in/15–50cm. This large fungus has numerous semicircular caps fused together into one large structure with a central stem. The caps are gray to brown, finely wooly-fibrous, fleshy, with the pores white, small and angular, descending the stem. The flesh is solid and white. **Spores** white, 5–7 x 3.5–5µm, smooth. At the base of old oaks and other deciduous trees, often where the tree has been lightning struck, widespread in North America. Edible and delicious. Specimens weighing up to 100lb (45kg) have been found.

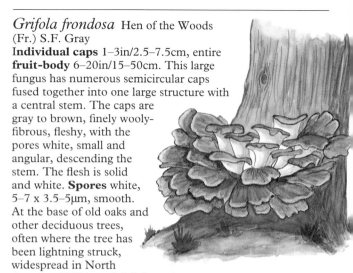

Meripilus giganteus
(Fr.) Kar.

10–30in/25–75cm across. This huge fungus has many broad shelves fused into a single mass. The shelves are fleshy, rather soft, yellow-ochre to tan, smooth to finely velvety with blunt margins. The pores are white. All parts bruise black. **Spores** white, 6–7 x 4.5–6µm, smooth. Growing at the base of dead or living deciduous trees, especially beech and oak, mainly in the east and south. Edible.

Laetiporus sulphureus Chicken Mushroom
(Fr) Murr.

Shelves 4–30in/10–75cm across.
Growing in large, overlapping masses,
each shelf is soft and fleshy, bright
orange-yellow to salmon above and
with lemon-yellow pores. **Spores**
white, 5–7 x 3.5–5µm, smooth.
Growing on dead or dying trees,
both deciduous and conifers it
can form huge clusters weighing
many pounds, often high up in
the tree but also at the base.
Edible and delicious and quite
unmistakable. There is a variety with
pinkish cap and white pores *semialbinus*
which is rather uncommon.

Postia caesia Blue-cheese Polypore
(Schr.) Karst. = *Tyromyces caesius*

Shelf 2–6in/5–15cm across. This
rounded, rather lumpy species is soft
and fleshy and starts grayish-white
but soon flushes pale blue
when rubbed or bruised.
The pores are small,
white and also bruise bluish.
The odor is fragrant and the
taste mild. **Spores** pale blue,
4–5 x 0.7–1µm, sausage-like
and smooth. Common on dead
wood throughout North America.

Hapalopilus nidulans
(Fr.) Kar.

Cap 1–5in/2.5–12.5cm. The deep
orange-brown fruit-body is fleshy,
thick, with small, cinnamon-brown
pores. The flesh is thick and watery,
tawny-brown. All parts turn bright
purple-violet with KOH solution.
Spores white, 3.5–5 x 2–3µm,
smooth. Common on fallen logs of
deciduous trees, mostly in the east.
Inedible.

Piptoporus betulinus Birch Polypore
(Fr.) Kar.

Cap 2–10in/5–25cm. The
rounded, kidney-shaped
caps are white to pale tan,
smooth, with an inrolled
margin. A short stem is
often present joining the
bracket to the tree. The
pores are extremely small
and white. **Spores** white,
5–6 x 1.5µm, sausage-
shaped. Found only on birch
trees, it is a serious parasite.
Inedible. The soft, white
flesh has a number of uses:
fire lighting, sharpening
razors, halting bleeding, and
today is cut in narrow strips
to pin insects in displays.

Trametes versicolor
(Fr.) Pilat

Cap 1–4in/2.5–10cm. Usually in overlapping clusters, each
bracket is thin, semicircular and with multicolored zones, from
brown to gray, blue-black, yellow to green. The fine pores are
pale yellowish. **Spores** white, 5–6 x 1.5–2.2µm, sausage-
shaped, smooth. Very common on fallen timber throughout
North America. Inedible.

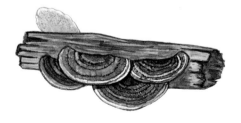

Daedaliopsis confragosa
(Fr.) Schroer.

Cap 2–6in/5–10cm. The semicircular to kidney-shaped brackets are tough, broad at the back with a thin margin, gray-brown and zoned. The pores vary from round to quite maze-like or in extreme forms even gill-like; they are pale cream-buff bruising pinkish. **Spores** white, 7–11 x 2–3µm, sausage-shaped, smooth. Very common on fallen timber, often persisting for several years, throughout most of North America. Inedible.

Daedalea quercina
Fries

Cap 2–6in/5–15cm. The thick, very tough bracket is smooth to rather cracked or furrowed on top, white to pale ochre, often slightly zonate. The pores are very elongated, maze-like with thick walls, whitish-buff. **Spores** white, 5.5–7 x 2.5–3.5µm, cylindrical, smooth. Frequent on dead oaks, from Maine down to North Carolina, west to Ohio. Inedible.

Ganoderma applanatum Artist's Fungus
(Pers. ex Wall.) Pat.

Cap 4–20in/10–50cm. The very hard, woody brackets are thick, tapering to the blunt margin, lasting for many years and adding tube layer to tube layer each year. The upper surface is smooth to lumpy, crust-like, gray-brown with a powdery covering of brown spores (which are attracted and held by a static charge formed on the surface). The margin is white and the pores on the underside are cream bruising instantly brown when scratched. The flesh is pale cinnamon with small whitish flecks. **Spores** reddish-brown, 6.5–9.5 x 5–7μm, with a thick double wall, perforated on the outer layer. Common on deciduous trees both living and dead, throughout North America. Scratching the pores with a toothpick, you can draw a picture which will become permanent as the fungus dries. The thicker, blunter *G. adspersum* has darker brown flesh without white flecks, and larger spores 8–13μm long.

Ganoderma tsugae
Murrill

Cap 4–12in/10–30cm. The kidney-shaped cap is deep purplish-red to brownish-orange and looks as if freshly lacquered; the margin is thick and white or orange. The tubes and pores are white.
Spores brown, 9–11 x 6–8μm, with thick, double walls. Common on dead conifers, especially hemlock, spruce and pine. The similar *G. lucidum* grows on oaks and chestnuts.

Fistulina hepatica Beefsteak Polypore
Schaeff. ex Fr.

Cap 3–10in/7.5–25cm. The tongue-like to semicircular bracket is soft and fleshy, gelatinous and often slimy on top, usually minutely roughened, pimpled, deep blood-red. The tubes are reddish with white pore mouths and are easily separable from each other. The flesh is fibrous, wet and exudes a reddish-brown liquid. **Spores** pinkish-salmon, 4.5–6 x 3–4µm, smooth. Uncommon, found on oak and chestnut trees in eastern North America. Edible but rather acidic, an acquired taste. The mushroom often drips "blood" and looks very liver- or tongue-like.

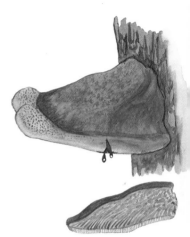

Fomitopsis pinicola
(Fr.) Kar.

Cap 3–10in/7.5–25cm. This hoof-like, hard and woody species has a dull brownish-red to ruby-red or even blackish crust which is usually furrowed and slightly zonate, redder near the margin. The pores are small, whitish bruising yellow. The flesh is thick, light yellowish-brown. The fungus adds a new tube layer each year. **Spores** white, 6–8 x 3.5–4µm, cylindrical, smooth. Common on dead conifers throughout most of North America.

Schizophyllum commune
Fries

Cap ⅜–2in/1–5cm. The small, thin caps are white and hairy. The undersurface looks at first like gills but are gill-like folds split lengthwise on the edges, pale pinkish-buff.
Spores white, 3–4 x 1–2.5µm, smooth. On fallen branches of deciduous trees, widespread. Throughout the world this species can be found all year, surviving droughts by rolling up the folds until wet weather returns. Inedible.

JELLY FUNGI AND RELATED GROUPS

All these fungi belong to a large group which share the common feature of having their spore-producing cells, the basidia, divided (septate) either longitudinally or transversely. They are usually of a jelly-like or rubbery consistency.

Tremella mesenterica Witches' Butter
Ret. ex Fr.
1–4in/2.5–10cm across. A common species throughout North America and throughout the year, whenever the weather is damp, it grows on twigs and fallen branches. It forms an irregular, lobed, jelly-like mass but when the weather turns dry it shrinks to a hard, horny orange lump, reviving when it rains again. **Spores** yellowish, 7–15 x 6–10µm. Basidia are divided longitudinally.

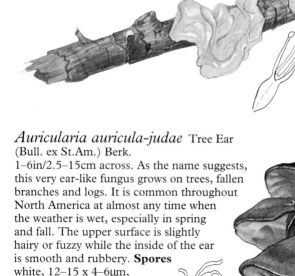

Auricularia auricula-judae Tree Ear
(Bull. ex St.Am.) Berk.
1–6in/2.5–15cm across. As the name suggests, this very ear-like fungus grows on trees, fallen branches and logs. It is common throughout North America at almost any time when the weather is wet, especially in spring and fall. The upper surface is slightly hairy or fuzzy while the inside of the ear is smooth and rubbery. **Spores** white, 12–15 x 4–6µm, sausage-like. Basidia are transversely septate. An edible species considered a delicacy.

Tremiscus helvelloides

(Pers.)Donk = *Phlogiotis helvelloides*
(Fr.) Mar.
1–4in/2.5–10cm across. A widely distributed but never very
common species, although most frequent in the Pacific
Northwest. The beautiful pink to apricot, often funnel
shaped, wavy or lobed fruit-bodies, are unmistakable.
They grow in clusters on the ground under
conifers or on very rotted wood in the fall. The flesh
is rubbery. **Spores** white, 9–12 x 4–6μm. Basidia are
longitudinally septate. Edible but rather tasteless.

Calocera viscosa

(Pers. ex Fr.) Fr.
1–4in/2.5–10cm high. This
coral-like mushroom is
bright golden-yellow and
has a tough, gelatinous
texture, unlike most of the
true Coral fungi. The
branches fork at the tips
and may be blunt to quite
pointed. **Spores** ochre-
yellow, 9–14 x 3–5μm,
sausage-shaped, smooth.
The basidia are shaped like
a tuning-fork. Common on
dead coniferous wood
throughout Much of North
America. Inedible. *C. cornea*
differs in its very small, simple,
unforked clubs which are pale yellow
and grows on deciduous wood.

ASCOMYCETES

CUP FUNGI AND
RELATED FUNGI

These fungi produce their spores in a quite different way from all the other fungi shown in this book. Their spores are contained in a special cell called an ascus, usually 8 spores per cell although this can vary depending on the genus and species concerned. The spores are ejected from the ascus like bullets from a gun, often producing a cloud of spores above the fungus when it is picked or disturbed in any way. The ascus gives its name to the group as a whole – the Ascomycetes.

CUP FUNGI AND RELATED FUNGI

Morchella esculenta Yellow Morel
L. ex Fries

Cap 2–5in/5–12.5cm high. These highly sought after mushrooms are very distinctive with their spongy caps set on swollen stems. The cap begins tightly compressed and dark blackish-brown with white ridges; as it expands it turns a bright yellow-ochre and the ridges become thin and jagged. The microscopic asci line the pits and ridges with their spores. The stem is white and very rough or granular, and is usually swollen at the base. The entire mushroom is hollow when cut in half. **Spores** (8 per ascus) are deep yellow-ochre in deposit, 20–24 x 12–24µm, elliptical, smooth. Found in the spring for a few weeks under dying apple, elm and ash trees. A wonderful edible species often growing in enormous numbers where conditions are right. Many different forms or species have been described, a small yellowish one is very common under tulip trees (*Liriodendron*) and a small gray one under ash.

Morchella elata Black Morel
Fries

Cap 2–5in/5–12.5cm high. A
rather conical, narrow species
with often very regular,
parallel ribs with cross ridges,
the cap is a dark smoky,
blackish-gray, sometimes paler
in the pits when mature. The
stem is usually cylindrical,
white and roughened. The cap
and stem are hollow. **Spores**
pale ochre, 24–28 x 12–14μm,
elliptical, smooth. Often
connected with conifers or
ash, found only in the spring,
widespread. Edible and
delicious. The Burn-site
Morel, *M. atromentosa*, differs
in both its dark cap and stem
being minutely blackish,
velvety and it grows on the site
of recent fires in woods in
central and western states.

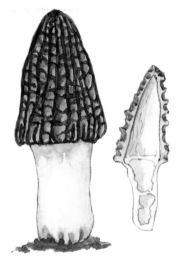

Morchella angusticeps
Peck

Cap 1–2in/2.5–5cm high.
This species has a small cap
in proportion to the long
stem and there is hardly any
division between the cap
and the stem. The cap is
conical, dark grayish-brown
with parallel ridges. The
flesh of both cap and stem
is rather thin unlike other
morels. The white stem
soon flushes pinkish-brown.
The entire fungus is hollow.
Spores pale ochre, 24–28 x
12–14μm, elliptical,
smooth. In mixed woods in
early spring, uncommon.
Edible but not as good as
other morels.

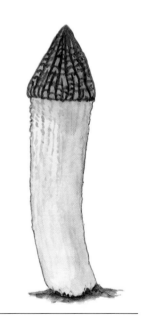

Verpa bohemica
(Kromb.) Schroet.

Cap 1–1¼in/2.5–3cm high. The thimble-like cap is very wrinkled and is not attached to the tall stem except at the very tip. The color is a dark yellow-brown. The stem is cylindrical, whitish, smooth or with granular ridges, and is hollow. **Spores** pale ochre, enormous in size 60–80 x 15–18µm, and only 2 are found in each ascus. Common in damp woods, streamsides in spring, throughout North America although apparently rare in eastern states. Edible in small quantities, but best avoided as it is known to interfere with muscular coordination.

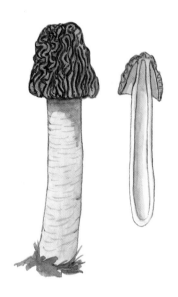

Verpa conica
Swartz ex Pers.

Cap ⅜–1¼in/1–3cm high. The thimble-shaped cap is smooth, not wrinkled, and only attached to the long stem at the very apex. The smooth to granular stem is white to yellow-ochre and completely hollow. **Spores** pale ochre, 22–26 x 12–16µm, elliptical, smooth, and 8 per ascus. Frequent in shady woodlands and old orchards, widespread. Edible but not very choice.

Morchella semilibera
DC. ex Fries

Cap 1–2in/2.5–5cm high. The conical to blunt cap is dark brown to yellow-brown with darker, almost parallel ridges and is attached to the stem for about half of its height. The stem is small at first but may soon expand to become very tall and swollen in relation to the cap, and is whitish with a granular texture. The mushroom is completely hollow. **Spores** pale ochre, 24–30 x 12–15μm, elliptical, smooth. Common and usually the first morel to appear, under mixed deciduous trees, early spring, widespread. Edible but rather thin-fleshed.

Helvella crispa
Scop. ex Fr.

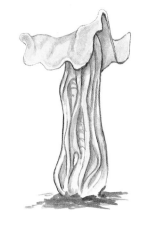

1–3in/2.5–7.5cm high. A beautiful, snow-white species with a curiously fluted, columned stem and saddle-shaped, thin cap. The whole fungus is very fragile and crisp, the stem is hollow with several convoluted chambers. **Spores** cream, 18–21 x 10–13μm, elliptical. Frequent in open, grassy woodlands, widespread, especially in the west. Edibility uncertain.

Helvella lacunosa
Afz. ex Fries
2–4in/5–10cm high. The entire fungus is an ashen, sooty gray-black with the stem fluted, twisted and ribbed, and the saddle-like thin cap often lobed and downcurved over the stem. The stem is hollow and chambered. **Spores** pale cream, 15–20 x 9–12µm, elliptical, smooth. Common in mixed woods, widespread but especially in the Pacific Northwest. Edibility doubtful.

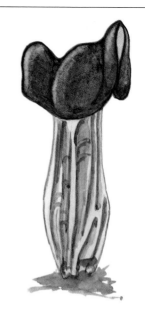

Gyromitra infula False Morel
(Schaeff. ex Fr.) Quel.
Cap 1–4in/2.5–10cm high. A dark reddish-brown, the saddle-like, thin cap clasps the rounded, cylindrical stem. The latter is flushed pale brown and is hollow with some chambers. **Spores** cream, 18–24 x 7µm, elliptical, smooth. Uncommon, on fallen conifer logs or debris, throughout North America. Poisonous. The very similar *G. ambigua* has violet tints in its brown cap and stem.

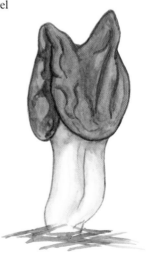

Gyromitra esculenta False Morel
(Pers. ex Fr.) Fr.

Cap 2–4in/5–10cm high. The rounded, very wrinkled, almost brain-like cap is a shiny reddish-brown, the interior is hollow and chambered. The stem is white to pale brown and smooth to granular. **Spores** cream, 18–22 x 9–1μm, elliptical, smooth. In the spring under conifers, widespread in the north and in mountainous areas. Poisonous and deadly, it contains a cumulative poison, a chemical similar to that used in rocket fuel.

Otidea onotica
(Pers.) Fuckel

Cup 1–2in/2.5–5cm. The long, rabbit ear-like cups are a bright apricot yellow, darker, more orange inside (which is lined with the microscopic asci). One or more "ears" may be fused together from a common base. **Spores** 12–14 x 5–6μm, elliptical, smooth. On soil and leaf-litter in mixed woods, uncommon, widespread. Edible. *O. leporina* is similar but a dull yellow-ochre.

Peziza vesiculosa
Bull.

Cup 1–3in/2.5–7.5cm. The yellow-ochre cup is paler, whitish and very rough, scurfy-granular on the outer surface. The edges of the cup are often jagged, tooth-like and incurved. **Spores** 20–24 x 12–14μm, elliptical, smooth. Not common but often in large groups on old compost, straw, manure heaps, widespread. Inedible.

Peziza phyllogena
Cooke = *P. badioconfusa*
Cup 1–4in/2.5–10cm. Found
in spring and early summer
this dark, reddish-brown cup
fungus is flushed with violet
on the outer surface and
often olive on the inside.
Spores 17–21 x 8–10μm,
elliptical, finely warted.
Common in mixed woods
on the ground, throughout North
America. Edible.

Aleuria aurantia Orange Peel
(Fr.) Fuckel
Cup 1–4in/2.5–10cm. The flattened cups really can be
mistaken for pieces of orange peel (and vice versa). The inner
surface is smooth and brilliant orange while the outer surface
is minutely scurfy-hairy and paler, whitish-orange. **Spores**
18–22 x 9–10μm, elliptical with a network of ridges. Common
in clusters on soil by paths, roads and disturbed areas,
widespread. Edible.

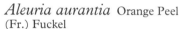

Scutellinia scutellata
(L. ex Fr.) Lamb.
Cup ¼–½in/0.5–1.2cm. The deep
scarlet cup is fringed with short,
black "eyelashes" all around the margin.
The undersurface of the cup is minutely hairy.
Spores 18–19 x 10–12μm,
elliptical, warty. Very common
on wet, rotten logs or soil,
throughout North America.
This is a complex of closely
related species differing mostly
microscopically.

Trichoglossum hirsutum Earth Tongue
(Pers.) Boud.

Club 1–3in/2.5–7.5cm high.
This all-black, flattened club
is minutely hairy all over (use
a hand lens). The club is often
grooved or twisted. **Spores**
enormously long and thin,
100–150 x 6–7μm, brown,
divided into 16 chambers.
Common, often in large
numbers on the forest floor or
in sphagnum moss, widely
distributed. Inedible. There
are many other species
differing in their spores and
the number of chambers
therein. The genus *Geoglossum*
looks identical but is
completely smooth.

Leotia lubrica Jelly Babies
Pers. ex Fr.

Club 1–2in/2.5–5cm high.
Forming a rounded,
gelatinous head on a slender,
rubbery stem. The cap is
ochre-yellow to olive-yellow
while the stem is yellow-
orange with darker, greenish
dots or granules. **Spores**
17–26 x 4–6μm, spindle-
shaped. Common, in small
clusters in leaf-litter
throughout North America.
L. viscosa has a green head
on a yellow stem and *L.
atrovirens* is entirely green,
but there is some evidence
that these are the result of
attack by an imperfect
fungus on ordinary
L. lubrica.

Chlorociboria aeruginascens
(Nyl.) Kan.
Cup ¼in/0.2cm. The thin, flattened cup is a vivid blue-green, almost unique amongst cup fungi. The mature cups are rather rare, and it is more likely the wood on which it grows will be seen stained with bright blue-green throughout its length as if dyed. **Spores** 6–10 x 1.5–2μm, spindle-shaped. Common on fallen branches, especially oak, throughout North America.

Galiella rufa Rubber Cup
(Schw.) Nannf. & Korf
Cup 1–2in/2.5–5cm. The deep reddish-brown, rubbery or gelatinous cups are thick, with a hairy outer surface and the margin inrolled over the smooth inner surface. The cup has a short stem. **Spores** 20 x 10μm, elliptical, pointed at the ends and with small warts. Common, often in groups on fallen branches and twigs of deciduous wood, eastern North America.

Xylaria hypoxylon
(L. ex Hook.) Grev.
Club 1–3in/2.5–7.5cm high. The thin, flattened, and often branched clubs look like blackened antlers with the base a sooty black but the tips white. The white powder is formed of asexual spores, later it will turn black. **Spores** 11–14 x 5–6μm, bean-shaped, black. Very common on dead wood, throughout North America.

Xylaria polymorpha Dead Man's Fingers
(Pers. ex Mer.) Grev.

Club 2–4in/5–10cm high. The swollen, irregular clubs are usually single but in groups from a common base; they are black and roughened, granular. They are sometimes white and powdery with asexual spores. The flesh is thick, white. **Spores** 20–30 x 5–10μm, spindle-shaped, black. Common on dead wood throughout North America.

Cordyceps ophioglossoides
(Ehr. ex Fr.) Link

Club 2–4in/5–10cm high. This slender club has a distinct, reddish-brown, granular head set on a yellow, smooth stem. If the stem is carefully followed down into the soil, bright golden threads will be found at the base and these are attached to another fungus, a false truffle *Elaphomyces*, a round, granular reddish-orange ball. **Spores** of *Cordyceps* are 2.5–5 x 2μm, elliptical, smooth. Quite common on the forest floor in eastern North America.

Appendix

Useful Chemicals

Ferrous sulphate (FeSO4) 10% solution in water, very important in a wide range of mushrooms, especiallly the genus *Russula*. The reaction varies from pink to gray, bluish or dark green.

Ammonia (NH3) 5% solution in water (houshold glass cleaners with ammonia are quite sufficient). Very useful, reactions range from bright red to violet or green.

Potassium Hydroxide (KOH) 10% solution in water, positive reactions are usually yellow to orange or reddish-brown.

Melzer's Solution Formula is: distilled water 20cc, potassium iodine 1g, iodine 0.5g, chloral hydrate 20g. Essential in the study of *Russula* and *Lactarius* and many other fungi if the microscope is to be used.

VERY IMPORTANT! These chemicals are poisonous and/or corrosive. Great care should be taken when handling.

Mushroom Societies

You should join your local mushroom group as soon as possible. For information on where to contact them, and to find out about mushrooming nationwide, contact the North American Mycological Association. Their address is N.A.M.A. Membership Secretary, 336 Lennox Ave, Oakland, CA. 94610-4675.

Index